Mathematical Theories of Populations: Demographics, Genetics and Epidemics

FRANK HOPPENSTAEDT
Courant Institute of Mathematical Sciences
New York University

Mathematical Theories of Populations: Demographics, Genetics and Epidemics

SOCIETY FOR INDUSTRIAL AND APPLIED MATHEMATICS
PHILADELPHIA

10 9 8 7 6 5 4

Library of Congress Cataloging-in-Publication Data

Hoppensteadt, F. C.
Mathematical theories of populations : demographics, genetics and epidemics / Frank Hoppensteadt.
p. cm. — (CBMS-NSF regional conference series in applied mathematics)
Includes bibliographical references (p.).
ISBN-10: 0-89871-017-0 (pbk.)
ISBN-13: 978-0-898710-17-5 (pbk.)
1. Population—Mathematical models. 2. Population genetics.
3. Epidemics. I. Title. II. Series
HB849.51.H66 1997
304.6'01'51—dc21 97-19735

Contents

Preface vii

I. THE EQUATIONS OF POPULATION DYNAMICS 1

1. Age dependent population growth 1
    1.1 Age independent version 3
    1.2. Solution for $u(a, t)$ 4
    1.3. Example: the genesis model 5
2. Analysis of the birth rate: stable age distribution 6
    2.1. Example. age independent case 6
    2.2. Analysis of the birth rate 7
3. A model of a self-limiting population 9
    3.1. An age independent version: the logistic equation 10
4. A two-sex model 11

Bibliography 14

II. DETERMINISTIC MODELS IN GENETICS 17

1. A brief introduction to Mendelian genetics 17
2 The one-locus, two-allele model 19
    2.1. The basic model 20
    2.2. Slow selection by death 22
    2.3. Lethal genes 26
    2.4. The De Finetti diagram 27
    2.5. Assortative mating and mutation 29
3. Age dependent population genetics 32
4. Propagation of a gene in a spatially distributed population 36
    4.1. The form of the progressing waves 37
    4.2. Stability of progressing waves: $c = c_{min}$ 39
    4.3. Stability of progressing waves: $c > c_{min}$ 40

Bibliography 43

III. THEORIES OF EPIDEMICS 45

1. General theory of contagious phenomena 45
    1.1. Some examples of epidemic theories 45
    1.2. A general age dependent theory 50
    1.3. Nonhomogeneous mixing: a model of spatial spread 52
2. Qualitative behavior of deterministic epidemics 54
    2.1. The threshold theorem 54
    2.2. Relapse-recovery model 61
    2.3. Models of spatial spread 64

Bibliography 71

Preface

Mathematical theories of populations have been derived and effectively used in many contexts in the last two hundred years They have appeared both implicitly and explicitly in many important studies of populations: human populations as well as populations of animals, cells and viruses.

Several features of populations can be analyzed. First, growth and age structure can be studied by considering birth and death forces acting to change them. In particular, the long time state of the population and its sensitivity to changes in birth and death schedules can be determined.

Another population phenomenon which is amenable to analysis is the way various individual traits are propagated from one generation to the next. An extensive theory for this has evolved from the first observations of inheritance by Gregor Mendel.

In addition, the spread of a contagious phenomenon, such as disease, rumor, fad or information, can be studied by means of mathematical analysis. Of particular importance in this area are studies of the dependence of contagion on parameters such as contact and quarantine rates.

Finally, the dynamics of several interacting populations can be analyzed. Theories of interaction have become useful with recent studies of ecological systems and economic and social structures.

The most direct approach to the study of populations is the collection and analysis of data. However, serious questions arise about how this should be done. For one thing, facilities can be swamped by even simple manipulation and analysis of vast amounts of data. In addition, there are the questions: Which data should be collected? Which data adequately describe the phenomenon, in particular which are sensitive indicators for detecting the presence of a phenomenon?

A study of the population's underlying structure is essential for answering these questions, and mathematical theories provide a systematic way for doing this. Many techniques for analyzing complicated physical problems can be applied to population problems. In addition, many new techniques peculiar to these problems must be developed. Several population problems will be analyzed here which illustrate these methods and techniques.

The monograph begins with a study of population age structure. A basic model is derived first, and it reappears frequently throughout the remainder. Various extensions and modifications of the basic model are then applied to several population phenomena, such as stable age distributions, self-limiting effects and two-sex populations.

The second part is devoted to population genetics, and it contains a summary of some of the most successful applications of mathematics in the biological sciences. Attention is focused on the derivation and analysis of a model for a one-locus, two-allele trait in a large randomly mating population Then extensions of the system are considered which account for more complicated social structure (assortative mating and migration) and for age structure. This part ends with a description of Fisher's model for the propagation of a gene in a spatially distributed population, and stable gene waves are shown to exist. A reason for the success of mathematical theories in genetics has been the wealth of precise data which can be collected. Unfortunately, this is not the case in the topics discussed in Parts I and III.

The final part, Part III, is concerned with the dynamics of contagious phenomena in a population These are studied in the context of epidemic diseases, but the same methods can be used to describe other phenomena such as rumors, fads and information as well as models for two interacting systems. Several classic examples are discussed first, then a general age dependent theory is formulated. However, the emphasis in Part III is placed on studies of qualitative properties of several typical models. First, a threshold theorem is derived for an age dependent epidemic, and then the long time behavior of solutions to a relapse-recovery model is determined. Finally, models for the spatial spread of contagion are derived and extensively discussed.

Ecological systems and other complicated interacting population phenomena are not discussed in the monograph. This is primarily because the scope of these applications is too broad for the present study. Therefore, the only work mentioned in this direction is that relating to the interaction of two populations, such as in the Volterra-Lotka theory, and this only because it is equivalent to a susceptible-infective interaction model.

All of the theories discussed in this monograph are deterministic. This restriction has been made so that a much broader range of population phenomena can be discussed than only those for which stochastic theories (i.e., theories which account for random fluctuations in population size, parameters, etc.) have been derived and studied. Many interesting and important studies have been made with the stochastic analogues to some of the models developed here. While these are not reported here, references are frequently made to appropriate literature An introduction to these studies and recent work is given by Ludwig (1974) (see Part III).

Many topics covered here are approached from an unusual point of view which resulted from discussions with several co-workers. In particular, contributions to the demographic topics by Professor J. B. Keller and to the epidemic section by Professors D. Ludwig and P. Waltman must be thankfully acknowledged The monograph is based on a series of lectures given at the National Science Foundation Regional Conference which was held at the University of West Florida in August 1974, arranged by the Conference Board of Mathematical Sciences. In addition, this conference provided many useful discussions of several topics, this was made possible through the especially hard work of Professor S. Shamma and Miss L. Jerrigan in organizing and running the conference

FRANK HOPPENSTEADT

I. The Equations of Population Dynamics

Mathematical methods have been used for a long time to describe the dynamics of populations, and many of these applications have had tremendous influence on human populations. Although such uses of mathematics date back at least to Fibonacci, who in 1202 proposed his famous sequence of integers to model the growth of a rabbit population, one of the first important applications was made by Malthus at the end of the eighteenth century. Malthus proposed that human populations have a constant natural growth rate. This simple model has had a profound effect, in particular, in various branches of biological sciences and economics Evidence of this model's influence is the tremendous amount of research and analysis of populations which it stimulated, notably Darwin's work on natural selection. Even today the model is somewhat controversial since it is still being invoked in political arguments. One very important feature of the model was its simplicity since this made it accessible to a great many people.

The model was altered by Verhulst (1838) to account for self-limiting features of populations. Finally, Lotka (1922) formulated the basis of an age dependent theory which is the focal point of this section. A more complete historical setting for population models is created by Cole (1954) whose paper is reproduced in the book by Hazen (1970).

As an introduction to the basic methods in population mathematics, a model of age dependent population growth is constructed first. It is based on the equation of age dependent population growth, which will be shown to play a unifying role in population mathematics, and later, in the study of contagious phenomena such as epidemic diseases. This important equation has arisen in several applications of mathematics, in particular, in the work of McKendrick (1926) and Kermack and McKendrick (1932, 1933), and it has been used to study a great many biological and physical phenomena. It will be a basic building block in all of the population theories developed here. After this derivation in § 1, the remaining sections study various aspects of age dependent growth such as the existence of stable age distributions and self-limiting populations.

1. Age dependent population growth. A model is constructed in this section which illustrates some of the basic results of population mathematics. In this model, the dynamics of a one-sex population are described by means of an age distribution function $u(a, t)$. This function specifies the age structure of the population at time t. In particular, the number of individuals in the age bracket (a_1, a_2)

is given by the integral

$$N \int_{a_1}^{a_2} u(a, t)\, da,$$

where N is some scaling factor, for example, the population's size at $t = 0$. $u(a, t)$ is taken to be a smooth function of (a, t). While it may seem to be, this is not necessarily a strong restriction for example, if N is a large number, a unit change in the population corresponds to a small change in u.

Two basic assumptions are now made about the population. These prescribe how individuals are removed from and introduced into the population The first specifies how individuals are removed.

ASSUMPTION A. The change occurring in the population of age a at time t in a time interval of length h is proportional to the size of the population and h. Thus,

$$u(a + h, t + h) - u(a, t) = -\mu(a, t)u(a, t)h.$$

$\mu(a, t)$ is called the age-specific death rate, and it is assumed to be nonnegative ($\mu \geqq 0$).

Dividing both sides of this equation by h and passing to the limit $h = 0$, we have

$$\text{(I.1)} \qquad \frac{\partial u}{\partial t} + \frac{\partial u}{\partial a} = -\mu(a, t)u.$$

This equation will be referred to as the *equation of age dependent population growth*. It is a first order partial differential equation whose characteristics are straight lines parallel to the line $t = a$ in Fig. I.1. Later we shall make use of the fact that this equation can be reduced to a solvable ordinary differential equation along these characteristic lines.

The second assumption specifies the relation between the birth rate and the age structure of the population:

ASSUMPTION B. The number of individuals introduced into the population in the time interval $(t, t + h)$ is

$$Nh \int_0^{\infty} \lambda(a, t)u(a, t)\, da.$$

The number $Nh\lambda(a, t)$ gives the births produced by an individual of age a in the time interval $(t, t + h)$, and λ is called the age specific fertility of the population.

It follows from Assumption B that the birth rate $u(0, t)$ satisfies

$$\text{(I.2)} \qquad u(0, t) = \int_0^{\infty} \lambda(a, t)u(a, t)\, da.$$

Finally, the initial age distribution of the population is assumed to be known,

$$\text{(I.3)} \qquad u(a, 0) = u_0(a),$$

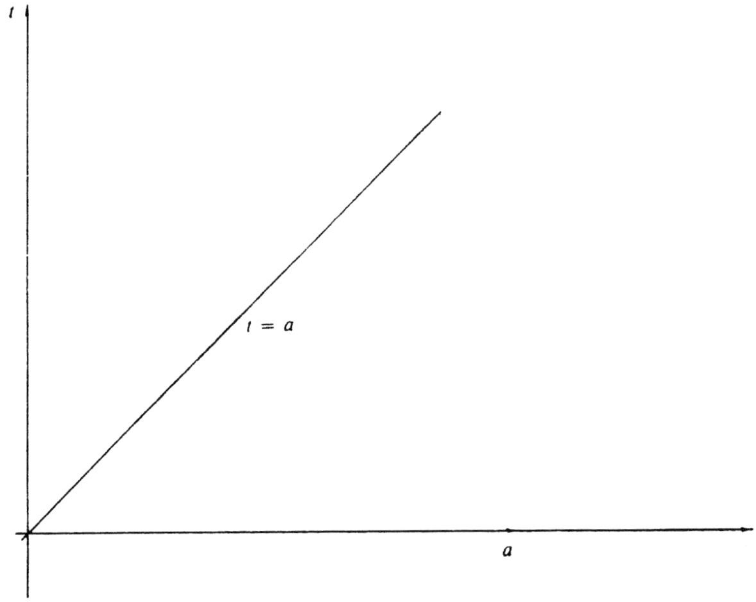

FIG 1.1

where u_0 is a smooth function. Typically, this function is zero for large a: for example, $a \geqq A$, A a fixed positive constant. If this is so, it can be shown that $u(a, t) = 0$ for $a \geqq A + t$ and so the integral in Assumption B and (I.2) is always over a finite interval.

The equations (I.1–3) make up the full model. If this can be solved for the function $u(a, t)$, then the dynamics of the population can be predicted. The model has appeared in various contexts in various forms. It was implicit in the work of McKendrick (1910, 1926, 1930), it arose in the work of von Foerster (1959) and Frederickson et al. (1963, 1967), as well as Rubinow (1968). Also, it has been studied in the demographic sense by Coale (1972) and Keyfitz (1968). A practical discrete time version was proposed by Leslie (1945, 1948). This, as well as a wealth of data, are given by Keyfitz and Flieger (1971).

1.1. Age independent version. Now, we investigate several methods for analyzing this model. The age independent version which occurs when λ and μ are independent of a is the most easily studied. In this case, we may consider the total population size at time t which is given by

$$U(t) = \int_0^\infty u(a, t)\, da$$

It follows that U satisfies

$$dU/dt = [\lambda(t) - \mu(t)]U(t), \qquad U(0) = \int_0^\infty u_0(a)\, da. \tag{I.4}$$

The solution of this problem is

$$U(t) = U(0)\exp\left[\int_0^t (\lambda(t') - \mu(t'))\,dt'\right]$$

and we see that the population grows or decreases depending upon the sign of $\lambda(t) - \mu(t)$ as our intuition would suggest The model (I.4) with $\lambda(t) - \mu(t)$ constant is the one proposed by Malthus, and the difference $\lambda - \mu$ is often called the biotic potential, intrinsic natural growth rate or malthusian parameter of the population

1.2. Solution for $u(a, t)$. Returning to the full problem (I.1–3), we are faced with the determination of $u(a, t)$ for $a, t \geqq 0$. This problem can be reduced to an integral equation for the birth rate. If the function $u(0, t) = B(t)$ were known, the solution could be found directly by introducing the characteristic coordinates and integrating the ordinary differential equations which result. In fact, the solution is

$$\text{(I.5)} \qquad u(a, t) = \begin{cases} u_0(a - t)\exp\left[-\int_0^t \mu(a - t + \sigma, \sigma)\,d\sigma\right] & \text{for } a \geqq t, \\ B(t - a)\exp\left[-\int_0^a \mu(\sigma, t - a + \sigma)\,d\sigma\right] & \text{for } t > a \end{cases}$$

However, B is defined by (I.2) which involves u. In particular, we have

$$B(t) = u(0, t) = \int_0^\infty \lambda(a, t)u(a, t)\,da.$$

Substitution of the result from (I.5) gives

$$B(t) = \int_0^t \lambda(a, t)B(t - a)k(a, t)\,da + \int_t^\infty \lambda(a, t)u_0(a - t)m(a, t)\,da,$$

where we have set

$$k(a, t) = \exp\left[-\int_0^a \mu(\sigma, t - a + \sigma)\,d\sigma\right]$$

and

$$m(a, t) = \exp\left[-\int_0^t \mu(a - t + \sigma, \sigma)\,d\sigma\right].$$

Rewriting this equation, we obtain an integral equation for B:

$$\text{(I.6)} \qquad B(t) = f(t) + \int_0^t K(a, t)B(t - a)\,da,$$

where we have set

$$f(t) = \int_t^\infty \lambda(a, t)m(a, t)u_0(a - t)\,da \quad \text{and} \quad K(a, t) = \lambda(a, t)k(a, t)$$

Both f and K involve only known data. If this equation can be solved for B, the solution of the problem (I.1–3) can be determined immediately from the equations (I.5). In this way, the full problem for $u(a, t)$ has been reduced to an integral equation for $u(0, t)$.

In the case where λ and μ are independent of t, the function K is also. Then (I.6) becomes

$$B(t) = f(t) + \int_0^t K(t - a)B(a)\,da, \tag{I.7}$$

which is called the *renewal equation*. There are several possible approaches to analyzing this equation: e.g., transform methods, successive approximations and various numerical methods. Detailed discussions of these can be found in Feller (1970) and Bellman and Cooke (1963). Of particular interest in the renewal theory, however, are the limit theorems developed for this equation. Using various conditions on f and K, one can obtain formulas which describe the behavior of B as $t \to \infty$. These results are derived in § 2.

1.3. Example: the genesis model. A particularly simple example, but yet one which illustrates the general approach outlined above, arises when the initial population distribution is concentrated at $a = 0$. To describe this situation, we make use of the Dirac delta function, $\delta(a)$, (see, for example, Lighthill (1958)), and define

$$u_0(a) = \delta(a),$$

which indicates that the number of individuals in the age bracket $[0, A)$ is $N\int_0^A u_0(a)\,da = N$ for any $A > 0$.

Let us also suppose that λ and μ are constant. The equation for B then becomes

$$B(t) = \lambda e^{-\mu t} + \lambda \int_0^t e^{-\mu a} B(t - a)\,da.$$

This equation can be solved for B. In fact, setting $\varphi(t) = e^{\mu t}B(t)$, we have

$$\varphi(t) = \lambda + \lambda \int_0^t \varphi(t - a)\,da \quad \text{or equivalently} \quad \varphi(t) = \lambda e^{\lambda t}.$$

Therefore, $B(t) = \lambda e^{(\lambda - \mu)t}$. With this, the solution of the full problem can be found as in (I.5):

$$u(a, t) = \begin{cases} \lambda \exp[(\lambda - \mu)(t - a) - \mu a] & \text{for} \quad t > a, \\ \delta(a - t)\exp[-\mu t] & \text{for} \quad a \geqq t. \end{cases}$$

In this case, the total population's dynamics are described by

$$U(t) = \int_0^\infty u(a, t)\,da = \exp[(\lambda - \mu)t],$$

where again the intrinsic natural growth rate appears

2. Analysis of the birth rate: stable age distribution. A question of practical importance has to do with the existence of a solution to (I.1–3) of the form

$$u = A(a)T(t).$$

If the solution of (I.1–3) approaches this form as $t \to \infty$, then the proportion of the population in any age bracket approaches a constant value as $t \to \infty$. In particular, the proportion of the population in the age bracket $[a_0, a_1]$ satisfies

$$\frac{\int_{a_0}^{a_1} u(a,t)\,da}{\int_0^\infty u(a,t)\,da} \to \frac{\int_{a_0}^{a_1} A(a)\,da}{\int_0^\infty A(a)\,da} \quad \text{as} \quad t \to \infty.$$

In this sense, the age distribution of the population is stable. To investigate the existence of such a stable age distribution, let us assume λ and μ are independent of t. Substitution of AT into (I.1) gives

$$(dT/dt)/T = -[(dA/da) + \mu A]/A = p = \text{const.},$$

so

$$u(a,t) = A(0)\exp\left[p(t-a) - \int_0^a \mu(a')\,da'\right],$$

where p is a parameter. The condition at $a = 0$ implies that

$$1 = \int_0^\infty \lambda(a)\exp\left[-pa - \int_0^a \mu(a')\,da'\right]da, \tag{I.8}$$

which therefore gives a relation between λ, μ and p, necessary for there to be a stable age distribution We note that if $\lambda(a)\exp[-\int_0^a \mu(a')\,da']$ is bounded and not identically zero, then there is a unique real solution for p. In fact, in that case the integral is a monotone decreasing function of p, ranging from infinity to zero Equation (I.8) was discovered by Lotka in 1922, and he called it the *characteristic equation.*

2.1. Example: age independent case. We first analyze the case where λ and μ are positive constants. Then condition (I.8) is satisfied by the unique value

$$p = \lambda - \mu.$$

The stable age distribution is

$$A(0)\exp[\lambda(t-a) - \mu t],$$

where only the constant $A(0)$ is not known. It can be shown that given any initial age distribution having bounded support, the resulting solution $u(a,t)$ approaches a stable age distribution with the same intrinsic rate of natural growth, $\lambda - \mu$. Instead of doing this for the particular example, we turn to a general analysis of the renewal equation.

2.2. Analysis of the birth rate. The birth rate $B(t)$ will be studied under the following restrictions:

Suppose that $\lambda(a)$ and $\mu(a)$ are nonnegative, piecewise continuous functions which satisfy $\lambda(a) \equiv 0$ for $a \geqq a^* > 0$, and if $k(a) = \lambda(a)\exp[-\int_0^a \mu(a')\,da']$, then $\int_0^\infty k(a)\,da > 0$. Also, suppose that the initial age distribution vanishes for large age, say $u_0(a) = 0$ for $a \geqq a^*$. It follows that $f(t)$ in the renewal equation vanishes for $t \geqq a^*$.

$B(t)$ satisfies the renewal equation

$$B(t) = f(t) + \int_0^t k(a)B(t-a)\,da. \tag{I.7}$$

Equations of this form are called convolution equations, and they are amenable to the method of Laplace transforms. In fact, the Laplace transform of $B(t)$ is defined by

$$L(B)(p) = \int_0^\infty e^{-pt}B(t)\,dt$$

for values of p for which the integral converges. The conditions given above ensure that the Laplace transforms of f and k exist. Therefore, for values for which the Laplace transform of B exists, we have from (I.7) that

$$L(B)(p) = L(f)(p) + L(k)(p)L(B)(p).$$

In particular, the convolution integral is converted by the Laplace transformation into a product of Laplace transforms. Thus, as long as $L(k)(p) \neq 1$, we have

$$L(B)(p) = L(f)(p)/[1 - L(k)(p)].$$

The equation

$$L(k)(p) \equiv \int_0^\infty e^{-pa}k(a)\,da = 1$$

is exactly Lotka's characteristic equation. The conditions listed above for $k(p)$ are sufficient to ensure that this equation has a unique real solution, say $p = p^*$, as discussed earlier.

Next, it follows that $L(B)(p)$ exists for $\operatorname{Re} p > p^*$, and defines an analytic function. This fact follows from the definition of $L(B)$ once we can guarantee that $B(t)$ grows no faster than an exponential function of t In fact, for p with $\operatorname{Re} p > p^*$, we have that

$$|e^{-pt}B(t)| \leqq |f(t)e^{-pt}| + \int_0^t |k(a)e^{-pa}|\,|e^{-p(t-a)}B(t-a)|\,da$$

from which it follows that

$$|e^{-pt}B(t)|\left[1 - \int_0^\infty k(a)e^{-a\operatorname{Re} p}\,da\right] \leqq \sup_{0 \leqq t < \infty} |f(t)e^{-pt}|.$$

Since $\operatorname{Re} p > p^*$, $\int_0^\infty k(a)\, e^{-a \operatorname{Re} p}\, da < 1$. This shows that

$$|B(t)| \leqq K\, e^{t \operatorname{Re} p} \left[1 - \int_0^\infty k(a)\, e^{-a \operatorname{Re} p}\, da\right]^{-1}.$$

Therefore, the Laplace transform of B, $L(B)(p)$ is defined as an analytic function of p in the half-plane $\operatorname{Re} p > p^*$

Now that the transform of B has been determined in terms of the parameters, the transform must be inverted by means of the inverse transform to recover $B(t)$. This is given by

$$B(t) = (1/2\pi i) \lim_{\tau \to \infty} \int_{\sigma - i\tau}^{\sigma + i\tau} e^{pt} L(f)(p)[1 - L(k)(p)]^{-1}\, dp,$$

where σ is any real number greater than p^*

This integral can be evaluated by the method of residues. First, observe that if $p' \neq p^*$ is a root of the characteristic equation, then

$$1 = |L(k)(p')| < \int_0^\infty e^{-a \operatorname{Re} p'} k(a)\, da,$$

and so $\operatorname{Re} p' < \operatorname{Re} p^*$. One result of this is that if the characteristic roots are all distinct, $B(t)$ can be determined as an infinite series of exponential terms:

$$B(t) = B_0\, e^{p^* t} + \sum_{n=0}^{\infty} B_n\, e^{p_n t},$$

where p_n are the characteristic roots, all satisfying $\operatorname{Re} p_n < p^*$. The coefficients in this expansion can be determined by calculating the residue of the integrand at the characteristic roots. It may happen that some of the complex roots are not simple, and then the problem becomes more difficult. Even in that case, we may still write

$$B(t) = B_0\, e^{p^* t} + o(e^{p^* t}),$$

where B_0 is the residue calculated at $p = p^*$:

$$\begin{aligned} B_0 &= \operatorname*{res}_{p = p^*} L(f)(p)[1 - L(k)(p)]^{-1} \\ &= \left[\int_0^\infty e^{-p^* t} f(t)\, dt\right] \Big/ \left[1 - \int_0^\infty k(a)\, e^{-p^* a} a\, da\right]. \end{aligned}$$

The error $o(e^{p^* t})$ indicates terms which approach zero as $t \to \infty$ when divided by $e^{p^* t}$. These terms are therefore negligible compared to $B_0\, e^{p^* t}$ for large time.

Finally, this analysis can be applied to show that the population approaches a stable age distribution. In fact, the function $v = e^{-p^*(t-a)} \exp[-\int_0^a \mu(a')\, da'] - B_0$ satisfies

$$\frac{\partial v}{\partial t} + \frac{\partial v}{\partial a} = 0, \qquad v(0, t) = u(0, t)\, e^{-p^* t} - B_0$$

Since $v(0, t) \to 0$ as $t \to \infty$ and $v(a, t)$ is constant along the characteristic lines, we have that for any fixed $\tilde{a} > 0$,

$$\max_{0 \leqq a \leqq \tilde{a}} |v(a, t)| \to 0 \quad \text{as} \quad t \to \infty .$$

In fact, given $\varepsilon > 0$, choose T so large that $|v(0, t)| < \varepsilon$ for $t \geqq T$. Then $|v(a, t)| < \varepsilon$ in the entire triangular region $0 \leqq a \leqq t - T$ provided $t \geqq T$

Therefore,

$$e^{-p^*t}u(a, t) \to B_0 \exp\left[p^*a - \int_0^a \mu(a')\, da'\right]$$

as $t \to \infty$, uniformly for a in any interval of the form $0 \leqq a \leqq \tilde{a}$. Another way of describing this result is that

$$u(a, t) \sim e^{p^*t}B_0 \exp\left[p^*a - \int_0^a \mu(a')\, da'\right]$$

as $t \to \infty$, where B_0 is given by the residue formula. The stable age distribution is therefore given by

$$A(a) = B_0 \exp\left[p^*a - \int_0^a \mu(a')\, da'\right].$$

The analysis of the renewal equation has attracted much attention beginning with Lotka (1922) and Feller (1941, 1970). The work in the population context is described by Coale (1972)

3. A model of a self-limiting population. In the model of § 1, Assumption B specified the fertility as a known function when in fact it should probably depend on the population's size Verhulst (1838) suggested this as a modification to Malthus' model (1.4), and he accounted for population dependent fertility in a simple way which will be described in this section. First, an age dependent version of the Verhulst model will be derived.

Let us assume now that the fertility has the form

$$\lambda(a, t) = \nu(a, t)\left[1 - \int_0^\infty \gamma(a, a', t)u(a', t)\, da'\right]_+ ,$$

where $(x)_+ = \max[0, x]$, i.e., $(x)_+ = 0$ if $x < 0$ and $(x)_+ = x$ if $x > 0$. The function ν here measures the age specific fertility and γ measures the reciprocal of some critical population density at which the individuals of age a' force those of age a to stop reproducing. Thus, for all (a, t) such that

$$\int_0^\infty \gamma(a, a', t)u(a', t)\, da' > 1,$$

we have $\lambda(a, t) = 0$.

With this new fertility function, the model (I.1–3) becomes

$$\frac{\partial u}{\partial t} + \frac{\partial u}{\partial a} = -\mu(a,t)u,$$

$$u(0,t) = \int_0^\infty \nu(a,t)u(a,t)\left[1 - \int_0^\infty \gamma(a,a',t)u(a',t)\,da'\right]_+ da,$$

$$u(a,0) = u_0(a).$$

This is now a nonlinear problem for u, and there are many difficult and unanswered questions about the behavior of its solutions.

3.1. An age independent version: the logistic equation. Some insight to this can be gained by studying the age independent version. Let us assume that the data μ, ν and γ are constant, and let us consider the total population at time t which is again described by the function

$$U(t) = \int_0^\infty u(a,t)\,da.$$

It follows from the full model that U satisfies the differential equation

$$dU/dt = -\mu U + \nu U[1 - \gamma U]_+$$

and the initial condition

$$U(0) = \int_0^\infty u_0(a)\,da.$$

This problem can be solved explicitly. Three solutions are described in Fig. I.2.

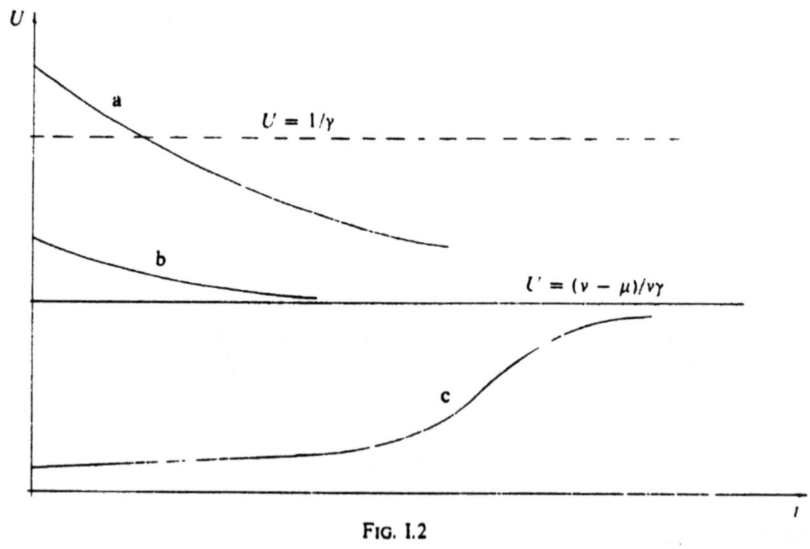

FIG. I.2

Verhulst, and many other investigators since, have offered evidence that the logistic equation gives a good model for the growth of certain populations. The main support of these theories comes from the observation that many biological populations exhibit sigmoid growth curves (c) However, there are many other equations whose solutions have the same shape, and some of these have also been offered as providing more realistic theories. Several of these are described and compared in the collection of papers by Hazen (1970). However, the logistic equation is certainly the most popular because of its simplicity, if not for its realism.

Let is now return to the age dependent model. One approach to analyzing it is by introducing the birth rate $B(t)$. As before, the solution of the problem is straightforward once B is known, and the problem can be reduced to an integral equation for B

$$B(t) = F(t) + \int_0^t k(a, t)B(t - a)\left[1 - g(t) - \int_0^t l(a, a', t)B(t - a')\,da'\right]_+ da.$$

Since we shall not attempt to analyze this general model further, the specific dependence of F, g, k and l on the data is omitted

A special case of interest arises when

$$\gamma(a, a', t) = \gamma\delta(a - a')$$

(δ is the Dirac delta function). This condition states that only those individuals of age a can influence their own fertility. In addition to this special form for γ, let us suppose the initial data are such that

$$\int_0^\infty \gamma(a, a', t)u_0(a')\,da' \equiv \gamma u_0(a) < 1$$

for all a This will ensure that $1 - \int_0^\infty \gamma(a, a', t)u(a', t)\,da' > 0$ for some initial time interval On this interval the equation takes the form

$$B(t) = F(t) + \int_0^t \nu(a, t)g(a, t)B(t - a)[1 - \gamma g(a, t)B(t - a)]\,da.$$

Even with this substantial simplification the model is difficult to analyze. A similar equation arises as an epidemic model, and the analysis of this kind of equation is postponed until Part III, § 2 2.

4. A two-sex model. A major defect of the model described in § 1 is that it involves only one function u to measure the size of one population. In using that model for human populations, u is usually taken to describe the female population. However, to get a more complete description of the population, both the females and the males must be accounted for This raises a question which has not yet been resolved: How does the coupling rate (or marriage rate) depend on the numbers of males and females? While the problem will not be resolved in this section, a sufficiently general approach will be taken which avoids this specific question.

We consider a two-sex population, and we let $m(a, t)$ and $f(a, t)$ denote the age distribution of males and females, respectively. As in § 1, equations for these functions may be derived from an assumption like A. We assume that

$$\frac{\partial m}{\partial t} + \frac{\partial m}{\partial a} = -\Delta_1(a, t)m,$$

$$\frac{\partial f}{\partial t} + \frac{\partial f}{\partial a} = -\Delta_2(a, t)f.$$

The functions Δ_1 and Δ_2 denote the removal rates for m and f Finally, the initial age distribution of each population is assumed to be known:

$$m(a, 0) = m_0(a), \quad f(a, 0) = f_0(a)$$

for $0 \leqq a < \infty$, and the birth rates are given by

$$m(0, t) = \int_0^\infty \int_0^\infty \lambda_1(a, a', t)c(a, a', t)\, da\, da',$$

$$f(0, t) = \int_0^\infty \int_0^\infty \lambda_2(a, a', t)c(a, a', t)\, da\, da',$$

where $c(a, a', t)$ gives the number of couples where the male has age a and the female age a', and λ_1 and λ_2 measure the productivity of such couples for male and female progeny, respectively.

The difficulty here is in determining c once m and f are known: First, the initial numbers of couples are assumed to be known,

$$c(a, a', 0) = c_0(a, a'),$$

and we assume the partners have positive ages,

$$c(a, 0, t) = c(0, a', t) = 0 \quad \text{for all } a, a' \text{ and } t$$

The dynamic equation for c is

$$\frac{\partial c}{\partial t} + \frac{\partial c}{\partial a} + \frac{\partial c}{\partial a'} = -[\Delta_1(a, t) + \Delta_2(a', t)]c - d(a, a', t)c + S(a, a', t; m_e, f_e).$$

This equation is obtained just as the equation for age dependent population growth was obtained in § 1 by balancing the change in the population over a short time interval with the forces acting in the population to change its structure. Now, a couple ceases to exist when one of the partners dies. Hence, the death rate is $\Delta_1 + \Delta_2$ The function $d(a, a', t)$ describes the divorce rate, and the functional S describes the source of couples It depends on the numbers of males and females eligible for coupling. These are given, respectively, by

$$m_e(a, t) = m(a, t) - \int_0^\infty c(a, a', t)\, da',$$

$$f_e(a', t) = f(a', t) - \int_0^\infty c(a, a', t)\, da.$$

The function S is called the *marriage function* by demographers, and they have considered its form extensively. Various choices have been made for this function, and then the predicted number of couples was compared with marriage data. For example, in the papers in Greville (1972) and Keyfitz (1972) several functions homogeneous of degree one in $m_e(a, t)$ and $f_e(a', t)$ were considered, such as the mean, the geometric mean, the harmonic mean and the minimum of the two functions. Unfortunately, the data available was unable to provide a clear choice between these functions. In fact, the marriage function should probably depend on the entire age distributions of males and females rather than just those of specific ages. Moreover, some thought suggests that this function may in fact be of degree two for small populations, then become of degree one as the populations increase. Without adequate data these questions cannot be resolved.

The cases considered by Keyfitz (1972) are summarized by taking S to be the following function of $m_e(a, t)$ and $f_e(a', t)$:

$$S = \mu(a, a', t)m_e(a, t)f_e(a', t)[m_e(a, t) + f_e(a', t)]^{-1}\rho(m_e(a, t)/f_e(a', t)),$$

where μ measures the rate at which formed couples "stick" (or marry), the harmonic mean is taken to denote the number of possible couplings between $m_e(a, t)$ and $f_e(a', t)$, and $\rho(m_e/f_e)$ is a function homogeneous of degree zero in m_e and f_e, which measures the adjustments needed to bias the harmonic mean. Since the source function should be proportional to the number of males when there are many females and to the number of females when there are many males, we assume $\rho(0) = \rho(\infty) = 1$.

Thus, the full model is a quite complicated system of nonlinear equations. From a practical point of view, the number of parameters must be reduced to a set which can be determined from available data, so the model is of limited use for quantitative work in its present form.

Some insight can be gained into the model by considering the special case where the males have a selective advantage over the females. Let us suppose that the data depend only on age, $\Delta_1 < \Delta_2$ and $\lambda_1 \geqq \lambda_2$. Then if the population is initially dominated by males, it will continue to be. It is reasonable in this case to take the coupling source function to be proportional to the number of females. Thus, we take $\rho(x) = 1 + x^{-1}$ for $x > 1$ and $\rho(x) = 1 + x$ for $x < 1$.

The dynamic equations then take on a simpler form:

$$\partial m/\partial t + \partial m/\partial a = -\Delta_1 m, \quad \partial f/\partial a + \partial f/\partial t = -\Delta_2 f,$$

$$\partial c/\partial t + \partial c/\partial a + \partial c/\partial a' = -(\Delta_1 + \Delta_2 + d)c + \mu f_e$$

Let us make the further assumption that an individual can mate only with someone his own age. To be precise, we assume

$$\mu(a, a') = \mu_0(a')\delta(a - a'),$$

where δ is the Dirac delta function described in § 1. Therefore, we expect

$$c(a, a', t) = g(a', t)\delta(a - a'),$$

where g remains to be determined.

Now since $f_e = f - g$, the equations for f and g become

$$\partial f/\partial a' + \partial f/\partial t = -\Delta_2 f,$$

$$\partial g/\partial a' + \partial g/\partial t = -[\Delta_1(a') + \Delta_2(a') + d]g + \mu_0(a')(f - g)$$

and

$$f(0,t) = \int_0^\infty \lambda_2(a)g(a,t)\,da, \quad g(0,t) = 0,$$

$$f(a',0) = f_0(a'), \quad g(a',0) = g_0(a').$$

Clearly, once g is determined, f can be found directly.

Finally, if the male death rate and the divorce rate are zero, then the problem can be reduced to an integral equation for the birth rate of females, $B(t) = f(0,t)$. In that case, the eligible females are given by f_e which satisfies

$$\partial f_e/\partial a + \partial f_e/\partial t = -(\Delta_2 + \mu_0)f_e,$$

$$f_e(0,t) = f(0,t), \quad f_e(a,0) = f_0(a) - g_0(a).$$

Thus, f_e can be found in terms of $B(t)$, and this function can be substituted into the birth rate equation for females to give a renewal equation for B This can be analyzed in the ways discussed earlier.

BIBLIOGRAPHY

R Bellman and K Cooke (1963) *Differential-Difference Equations*, Academic Press, New York

A. J Coale (1972) *The Growth and Structure of Human Populations, a Mathematical Investigation*, Princeton University Press, Princeton, N J

L. Fibonacci (1202) *Liber abbaci di Leonardo Pisano publicati da Baldasani Boncompagni*, Tipographic delle Scienze Math e Fisiche, Romalla

W Feller (1941) *On the integral equation of renewal theory*, Ann. Math Statist , 12, pp 243–267

——— (1970) *Introduction to Probability Theory and Applications*, vol II, John Wiley, New York

A G Fredrickson, D Ramkrishna and H M. Tsuchiya (1967) *Statistics and dynamics of procaryotic populations*, Math Biosci , 1, pp 327–374

A G Fredrickson and H M Tsuchiya (1963) *Continuous propagation of micro-organisms*, Amer Inst Chem Eng J , 9, pp 459–468

T N E Greville (ed) (1973) *Population Dynamics*, University of Wisconsin Press, Madison, Wisc

W E Hazen (ed) (1970) *Readings in Population and Community Ecology*, W B Saunders, Philadelphia

S Karlin (1966) *A First Course in Stochastic Processes* Academic Press, New York

W O Kermack and A G McKendrick (1932), Proc Roy Soc Ser A, 138, pp 55–83 (1933) Ibid , 141, pp 94–122

N Keyfitz (1968) *Introduction to the Mathematics of Populations*, Addison-Wesley, Reading, Mass

——— (1971) *What mathematical demography tells that we would not know without it*, in press

——— (1972) *The mathematics of sex and marriage*, 6th Berkeley Symp Math Stat Prob , Biology-Health section, Part II

N Keyfitz and W Flieger (1971) *Populations, Facts and Methods of Demography*, W H Freeman, San Francisco

P H Leslie (1945) *On the use of matrices in certain population mathematics*, Biometrika, 33, pp 183–212

——— (1948) *Some further notes on the use of matrices in population mathematics*, Biometrika, 35, pp 213–245

M. J Lighthill (1958) *Introduction to Fourier Analysis and Generalized Functions*, Cambridge University Press, London

A. J. LOTKA (1922) *The stability of the normal age distribution*, Proc. Nat. Acad. Sci., 8, pp. 339–345.

——— (1956) *Elements of Mathematical Biology*, Dover, New York.

R. M. MAY (1973) *Stability and Complexity in Model Ecosystems*, Princeton University Press, Princeton, N.J.

A. G. MCKENDRICK AND M. K. PAI (1910) *The rate of multiplication of micro-organisms. A mathematical study*, Proc. Roy. Soc. Edinburgh, 31, pp. 649–655.

A. G. MCKENDRICK (1926) *Applications of mathematics to medical problems*, Proc. Edinburgh Math. Soc., 44, pp. 98–130.

S. RUBINOW (1968) *A maturity-time representation for cell populations*, Biophysical J., 8, pp. 1055–1073.

S. RUSHTON AND A. J. MAUTNER (1955) *Deterministic model of a simple epidemic*, Biometrika, 42, pp. 126–132.

P. VERHULST (1838) *Notice sur la loi que la population suit dans son accroissement*, Corres. Math. et Phys. publié par A. Quetelet (Bruxelles), pp. 113–121.

II. Deterministic Models in Genetics

1. A brief introduction to Mendelian genetics. Population geneticists attempt to describe how populations of organisms reproduce and thereby propagate various properties to their progeny. Mendel's laws of inheritance serve as the basis for much of this work.

All organisms are made up of cells and they grow through cell reproduction. Cell reproduction occurs through *mitosis*, asexual reproduction where the cell doubles in size and then divides, and through *meiosis*, sexual reproduction where two cells interact to form a new cell. In meiosis, the parent cells create new cells, called *gametes*: one gamete from one parent then fuses with a gamete from the other parent to form a fertile cell, called a *zygote*. In humans, the gametes are the spermatoza (male) and ova (female). After a sperm and an egg fuse, the zygote goes on to reproduce through mitosis to form the embryo.

Cell reproduction therefore plays the central role in the study of genetics. A typical cell is made up of a *nucleus* which is surrounded by a mass, called the *cytoplasm*. This is all enclosed in the cell's membrane. Inside the nucleus are long thin string-like structures called *chromosomes*. These have very complicated chemical structures, composed like a chain whose links are made up of deoxyribonucleic acid (DNA), ribonucleic acid (RNA) and proteins. Biologists have shown that the chromosomes are the information carrying portion of the cell.

Some cells have single chromosomes, others have many matched sets. Cells which have isolated, distinct chromosomes are called *haploids*, those whose chromosomes occur in matched, homologous pairs are called *diploids*. Humans are diploid organisms having twenty-three pairs of chromosomes. The models studied in this chapter are for diploid organisms, but corresponding, and usually simpler, models can be constructed for haploids

Certain well-defined locations on the chromosomes are known to govern certain properties of the organism. The collection of chemicals at these locations are called *genes*, and a gene's location is called its *locus*. It is not known exactly how genes act to influence the organism, and this question is the centre of much current research. However, a great deal of phenomenologic evidence is available linking various genes with various properties of organisms. At one locus, the known gene appears in at least one variant form. The variant forms of the gene are called *alleles*. A gene may have two alleles or many, though the models constructed here are exclusively for two-allele genes.

When one locus with two alleles is considered, it is convenient to denote the alleles by A and a. In a diploid cell, there are three possible gene configurations at such a locus: AA, Aa, aa. These are called the *genotypes* of the locus. The two genotypes AA and aa are called *homozygotes* and the other genotype Aa is called the *heterozygote*.

It may be that the genotypes manifest themselves in distinct ways in the organism, as, for example, in the pelt coloring of certain foxes where the genotypes correspond to red, mottled and black pelts Or, one of the alleles may be *dominant*, such as in eye coloring in humans where the genotypes AA and Aa have the same eye color, but different from that of the genotype aa. In this case, A is called the *dominant* gene and a the *recessive* gene. This is also described by saying that AA and Aa have the *phenotype* A while aa has phenotype a.

Sexual reproduction occurs in diploid organisms through the formation of gametes which then fuse in meiosis to form zygotes. The gametes are haploid cells whose chromosomes are formed by one of the homologous strands of the parent cell. Thus, a genotype Aa creates gametes of genotype A and a, and all the gametes of a homozygote are identical.

Now, the basics of Mendelian genetics are easy to describe. When a genotype AA mates with a genotype Aa, what are the possible genotypes of the progeny? If half of the gametes carried by the heterozygotes are a and half A, then since the progeny must be either AA or Aa, half the progeny should be AA and half Aa. Similarly, in matings between two heterozygotes, one fourth of the progeny will be AA, one half Aa and one fourth aa. This is because only a formal distinction can be made between Aa and aA genotypes, and they are both counted as being Aa. Obviously, matings between identical homozygotes can produce only homozygous progeny. Finally, matings between an AA and an aa can result only in heterozygous progeny.

These results are summarized in Table 1.

TABLE 1
Mendel's laws of inheritance

Type of mating	Frequency of progeny by genotypes AA	Aa	aa
AA × AA	1	—	—
AA × Aa	1/2	1/2	—
Aa × Aa	1/4	1/2	1/4
Aa × aa	—	1/2	1/2
aa × aa	—	—	1
AA × aa	—	1	—

These laws play a fundamental role in the theory constructed here. Several outstanding introductions to this topic are available, particularly valuable are the books by Bonner and Mills (1964) and Lerner (1968). There are also several good treatments of modelling approaches taken here. Two which contain many good problems are Crow and Kimura (1970) and Moran (1962). An excellent book which contains much data was written by Wallace (1968). The books by Haldane (1932), Fisher (1958), S. Wright (1968, 1969) and Cavalli-Sforza and Bodmer (1971) must be recommended as source books in the area.

There is an extensive literature and great interest in the area of medical genetics. Excellent introductions to these topics are contained in the books by Harris (1966) and McKusick (1964). Harris' book requires some technical chemical background, but the McKusick book is quite accessible. A particularly important and interesting aspect of human genetics is the study of blood types. These two references give a background to the topic, as do Cavalli-Sforza and Bodmer.

2. The one-locus, two-allele model. The basic model derived in this section is for a large randomly mating diploid population. The goal of the model is to describe the dynamics of the genotype subpopulations determined by one locus having two alleles It is assumed that only birth and death are acting in the population to change its genotype distributions. A later section will describe the modifications necessary to account for migration, mutation and assortative mating.

The dynamics of the genotype subpopulations will be modelled by introducing functions to keep track of their sizes. Let the total population size be denoted by

$$Nn(t) = \text{total population size at time } t,$$

where N is some (large) scaling factor such as the population's initial size. We assume that $n(t)$ is a smoothly varying function of t. A Malthusian model will be assumed for the population where different birth and death rates are prescribed for the various genotype subpopulations.

The genotype subpopulations will be described by functions

$$\begin{aligned} Nn_{11}(t) &= \text{number of } A_1A_1 (= AA) \text{ genotypes at time } t, \\ Nn_{12}(t) &= \text{number of } A_1A_2 (= Aa) \text{ genotypes at time } t, \\ Nn_{22}(t) &= \text{number of } A_2A_2 (= aa) \text{ genotypes at time } t. \end{aligned}$$

These functions are also assumed to be smoothly varying functions of t.

The functions n, n_{11}, n_{12} and n_{22} will enter into the discussion only at the prelimary stages of the derivation. It is preferable to deal with frequencies rather than numbers, and so we introduce the functions

$$\begin{aligned} D(t) &= n_{11}(t)/n(t) = \text{proportion of population which is AA}, \\ 2H(t) &= n_{12}(t)/n(t) = \text{proportion of population which is Aa}, \\ R(t) &= n_{22}(t)/n(t) = \text{proportion of population which is aa}. \end{aligned}$$

The notation D, H, R is introduced for AA, Aa, aa, respectively, because the A gene may be taken as dominant and the a as recessive. However, no assumptions are made at this point about dominance of A and a.

Obviously, the identity

$$D(t) + 2H(t) + R(t) = 1$$

holds for all t.

The gene population carried by the locus in the organism population has $2Nn(t)$ members since each organism carries two genes. The proportion of the gene population which is of type A is given by

$$p(t) = (2n_{11}(t) + n_{12}(t))/2n(t) = D(t) + H(t)$$

since each AA carries two A genes and each Aa carries one. Similarly, the frequency of a genes in the gene population is

$$q(t) = (2n_{22}(t) + n_{12}(t))/2n(t) = R(t) + H(t),$$

and we have the identity

$$p(t) + q(t) = 1$$

for all t.

2.1. The basic model. Each genotype population is assumed to have fixed constant birth and death rates. These are denoted by

$$\begin{aligned} b_{ij} &= \text{birth rate of genotype } A_iA_j, \\ d_{ij} &= \text{death rate of genotype } A_iA_j, \end{aligned} \qquad i, j = 1, 2,$$

where $A_1 = A$ and $A_2 = a$. We take $b_{12} = b_{21}$ and $d_{12} = d_{21}$ since no distinction can be made between the genotypes A_1A_2 and A_2A_1.

The model is derived by balancing the changes which occur in the population with the forces acting to change it. In an interval of length h, the total population change is given by

$$N[n(t + h) - n(t)].$$

The changes are due to births and deaths. The number of births to AA genotypes in this interval is given by

$$hb_{11}Nn_{11}$$

and the number of deaths is

$$hd_{11}Nn_{11};$$

similarly for the Aa and aa genotypes. Thus,

$$n(t + h) - n(t) = h[(b_{11} - d_{11})n_{11} + (b_{12} - d_{12})n_{12} + (b_{22} - d_{22})n_{22}].$$

Dividing by h and letting $h \to 0$, we have

$$dn/dt = \overline{m}(t)n, \tag{II.1}$$

where

$$\overline{m}(t) = (b_{11} - d_{11})D(t) + 2(b_{12} - d_{12})H(t) + (b_{22} - d_{22})R(t).$$

Of course, the total population cannot be determined from this equation since its growth rate depends on the unknowns D, H and R.

Equations similar to (II.1) can be derived for n_{11}, n_{12} and n_{22}. The change in n_{11} over an interval of length h,

$$n_{11}(t + h) - n_{11}(t),$$

is due to births and deaths of AA genotypes The deaths occurring are measured by

$$hd_{11}n_{11}(t)$$

The number of births is not so easy to determine The number of births to AA females is $hb_{11}n_{11}$, however only a certain fraction of these, depending on who the mates were, will be AA genotypes Our assumption that the population is randomly mating comes into the model now in the following way The population is considered as being split exactly between males and females, each population having the same proportion of genotypes Then the number of AA females who will mate with an AA male is $n_{11}D$, with an Aa male is $2n_{11}H$ Therefore, the total AA progeny of AA females is determined by Mendel's laws as being

$$hb_{11}n_{11}(t)[D(t) + H(t)]$$

Heterozygous females may also bear AA progeny. These will be the result of matings with AA and Aa males Thus, the number of AA progeny of Aa females is, according to Mendel's laws,

$$hb_{12}n_{12}(t)[D(t) + H(t)]/2.$$

Now, we have accounted for all births of AA genotypes, and we can write

$$n_{11}(t+h) - n_{11}(t) = hb_{11}n_{11}(t)[D(t) + H(t)] + hb_{12}n_{12}(t)[D(t) + H(t)]/2 - d_{11}n_{11}(t).$$

Dividing by h and letting $h \to 0$, we then obtain a differential equation for n_{11}. Similar arguments must be applied to n_{12} and n_{22}. The differential equations which result from these calculations are

$$\begin{aligned} dn_{11}/dt &= b_{11}n_{11}[D + H] + b_{12}n_{12}[D + H]/2 - d_{11}n_{11}, \\ \text{(II.2)} \quad dn_{12}/dt &= b_{11}n_{11}[H + R] + b_{12}n_{12}/2 + b_{22}n_{22}[D + H] - d_{12}n_{12}, \\ dn_{22}/dt &= b_{12}n_{12}[H + R]/2 + b_{22}n_{22}[H + R] - d_{22}n_{22} \end{aligned}$$

Equations for the genotype frequencies can easily be obtained since

$$dD/dt = (d/dt)(n_{11}/n) = \frac{(dn_{11}/dt)}{n} - n_{11}\frac{(dn/dt)}{n^2},$$

similarly for H and R. The results of substituting these formulas into the equations in (II.1) and (II.2) are

$$\begin{aligned} dD/dt &= b_{11}Dp + b_{12}Hp - d_{11}D - \overline{m}D, \\ \text{(II 3)} \quad 2dH/dt &= b_{11}Dq + b_{12}H + b_{22}Rp - 2d_{12}H - 2\overline{m}H, \\ dR/dt &= b_{12}Hq + b_{22}Rq - d_{22}R - \overline{m}R, \end{aligned}$$

where we have used the gene frequencies $p = D + H$, $q = H + R$, to shorten the notation. Equations for the gene frequencies can be obtained directly from (II.3) as

$$\text{(II.4)}\qquad \begin{aligned} dp/dt &= p[m_1 - \bar{m}], \\ dq/dt &= q[m_2 - \bar{m}], \end{aligned}$$

where

$$\begin{aligned} m_1 = {} & \tfrac{1}{2}(b_{11} - b_{22})D + (b_{12} - b_{22})H + [D(b_{11} + b_{22}) + H(b_{12} + b_{12})]/2p \\ & - [d_{11}D + d_{12}H]/p, \end{aligned}$$

and

$$\begin{aligned} m_2 = {} & \tfrac{1}{2}(b_{22} - b_{11})R + (b_{12} - b_{11})H + [R(b_{11} + b_{22}) + H(b_{12} + b_{22})]/2q \\ & - [d_{22}R + d_{12}H]/q. \end{aligned}$$

Usually, the numbers $m_{ij} = b_{ij} - d_{ij}$ are called the fitnesses of the various genotypes, $\bar{m}$ is called the mean fitness of the population, and the complicated expressions for m_1 and m_2 are related to the fitness of the alleles A and a, respectively.

The equations (II.3) and (II.4) make up the basic one-locus, two-allele model, and we shall consider in the next several paragraphs various special cases and extensions of this model.

It should be emphasized that among the five variables D, H, R, p and q we know already two relations: namely,

$$D + 2H + R = 1 \quad \text{and} \quad p + q = 1 .$$

Moreover, p and q are defined in terms of D, H and R. Thus, the system could be reduced to two equations. However, this degeneracy will be maintained, and we shall deal with the full system, using the identities when convenient.

Once initial conditions for the genotypes are given, $D(0)$, $2H(0)$, $R(0)$, these equations can (in principle) be solved and the dynamics of the genotype and gene frequencies so determined.

2.2. Slow selection by death. Suppose the death rates of the various genotypes are almost equal. What effect will the resulting "slow selection" have on the gene frequencies? It is shown in this section how a perturbation method can be used to obtain an answer to this question.

The assumption is made explicit by setting

$$b_{ij} = b \quad \text{and} \quad d_{ij} = d + \varepsilon\Delta_{ij} \quad \text{for} \quad i, j = 1, 2,$$

where b, d and Δ_{ij} are fixed constants and ε is viewed as being a positive number very near zero. On setting $\varepsilon = 0$, we see that both the birth and death rates of all genotypes are the same, and then there is no selection in the model.

Putting these new birth and death rates into the basic model leads to the system

$$
\text{(II.5)}\quad
\begin{aligned}
dD/dt &= b(p^2 - D) + \varepsilon[D(\Delta_{11}D + 2\Delta_{12}H + \Delta_{22}R) - \Delta_{11}D],\\
2dH/dt &= 2b(pq - H) + \varepsilon[2H(\Delta_{11}D + 2\Delta_{12}H + \Delta_{22}R) - 2\Delta_{12}H],\\
dR/dt &= b(q^2 - R) + \varepsilon[R(\Delta_{11}D + 2\Delta_{12}H + \Delta_{22}R) - \Delta_{22}R],\\
dp/dt &= \varepsilon p[\Delta_{11}D + 2\Delta_{12}H + \Delta_{22}R - (\Delta_{11}D + \Delta_{12}H)/p].
\end{aligned}
$$

The equation for q will not be needed from here since we shall simply use the identity

$$q = 1 - p.$$

Although this appears to be a quite complicated system of equations, it is remarkably easy to obtain approximate solutions which are valid for all t when ε is small. Our approach involves the use of a powerful analytical method which takes advantage of the smallness of ε. The method is called the method of matched asymptotic expansions, and it is described in the paper by Hoppensteadt (1971). Unfortunately, the problem cannot be approached directly by simply setting $\varepsilon = 0$ in (II.5) and declaring that the solution of the no selection model gives a valid approximation to the solution of the slow selection model (II.5). It does give a valid approximation on fixed *finite* intervals, but not on the whole interval, $0 \leqq t < \infty$. Therefore, we take a more sophisticated approach.

First, we note that dp/dt is proportional to ε. Hence, p is expected to be a slowly changing function of t (its derivative is small). We account for this slow variation first by introducing a new slow time variable $\sigma = \varepsilon t$. The result is the system

$$
\begin{aligned}
\varepsilon dD/d\sigma &= b(p^2 - D) + \varepsilon\{\cdots\},^{1}\\
2\varepsilon dH/d\sigma &= 2b(pq - H) + \varepsilon\{\cdots\},\\
\varepsilon dR/d\sigma &= b(q^2 - R) + \varepsilon\{\cdots\},\\
dp/d\sigma &= p[\Delta_{11}D + 2\Delta_{12}H + \Delta_{22}R - (\Delta_{11}D + \Delta_{12}H)/p]
\end{aligned}
$$

We attempt to determine D, H, R and p as a power series in ε about $\varepsilon = 0$. Setting $\varepsilon = 0$ in this system, we get equations for the leading coefficients of these expansions:

$$D_0 = p_0^2, \quad H_0 = p_0q_0, \quad R_0 = q_0^2,$$

and

$$\text{(II.6)}\qquad dp_0/d\sigma = -p_0(1 - p_0)[2Ep_0 - (\Delta_{22} - \Delta_{12})],$$

where

$$2E = \Delta_{11} + \Delta_{22} - 2\Delta_{12}.$$

This equation has three constant solutions: $p_0 = 0$, $p_0 = 1$ and the third determined from the third factor on the right side, which may be admissible (i.e., satisfy

[1] The notation $\varepsilon\{\ \}$ denotes terms which are multiplied by ε and not significant in these calculations.

$0 \leqq p_0 \leqq 1$) depending on the particular choice of the Δ_{ij}. The solution of this equation exists for all $0 \leqq \sigma$, satisfies $0 \leqq p_0(\sigma) \leqq 1$, and approaches one of the constant solutions. The solution is uniquely determined by the initial condition

$$p_0(0) = D(0) + H(0),$$

the initial frequency of the A allele.

The next step in the analysis is to subtract off the slow time solution by defining

$$\sigma = \varepsilon t, \quad \tilde{D} = D - D_0(\varepsilon t), \quad \tilde{H} = H - H_0(\varepsilon t),$$
$$\tilde{R} = R - R_0(\varepsilon t), \quad \tilde{p} = p - p_0(\varepsilon t).$$

These functions satisfy the system

$$\begin{aligned} d\tilde{D}/dt &= -b\tilde{D} + \varepsilon\{\cdots\}, \\ d\tilde{H}/dt &= -b\tilde{H} + \varepsilon\{\cdots\}, \\ d\tilde{R}/dt &= -b\tilde{R} + \varepsilon\{\cdots\}, \\ d\tilde{p}/dt &= \varepsilon\{\cdots\}. \end{aligned}$$

This system is to be solved subject to the initial conditions

$$\begin{aligned} \tilde{D}(0) &= D(0) - D_0(0), \\ \tilde{H}(0) &= H(0) - H_0(0), \\ \tilde{R}(0) &= R(0) - R_0(0), \\ \tilde{p}(0) &= p(0) - p_0(0) = 0. \end{aligned}$$

Again the solution is to be determined as a power series in ε. The leading coefficients in these expansions are found by solving this system with $\varepsilon = 0$. Therefore,

$$\tilde{D}_0(t) = e^{-bt}\tilde{D}_0(0), \quad \tilde{H}_0(t) = e^{-bt}\tilde{H}_0(0), \quad \tilde{R}_0(t) = e^{-bt}\tilde{R}_0(0), \quad \tilde{p}_0(t) = 0.$$

This can be combined with the solution of the slow time problem to give that the solution of (II.5) satisfies

$$\begin{aligned} D(t,\varepsilon) &= p_0^2(\varepsilon t) + e^{-bt}[D(0) - p_0^2(0)] + O(\varepsilon), \\ H(t,\varepsilon) &= p_0(\varepsilon t)q_0(\varepsilon t) + e^{-bt}[H(0) - p_0(0)q_0(0)] + O(\varepsilon), \\ R(t,\varepsilon) &= q_0^2(\varepsilon t) + e^{-bt}[R(0) - q_0^2(0)] + O(\varepsilon), \\ p(t,\varepsilon) &= p_0(\varepsilon t) + O(\varepsilon), \end{aligned}$$

where p_0 is determined by solving (II.6) and $q_0 = 1 - p_0$. The notation $O(\varepsilon)$ denotes errors that approach zero as $\varepsilon \to 0$ uniformly on any interval of the form $0 \leqq t \leqq T/\varepsilon$. T may be taken as ∞ if it is known that p_0 approaches a constant at an exponential rate.

This approximation shows that for small ε, the genotype distributions converge rapidly to the distributions

$$p_0^2(\sigma), \quad 2p_0(\sigma)q_0(\sigma), \quad q_0^2(\sigma).$$

These are called the Hardy-Weinberg proportions, and in discrete time models with no selection, these are attained after only one generation. However, here the Hardy-Weinberg proportions are slowly changing to whatever values determined from (II.6).

With a little more work, higher order approximations, up to order $O(\varepsilon^2)$ can be obtained from the perturbation method.

Several special cases of (II.6) are of practical interest and can be analyzed quite easily:

(a) *No epistasis*. $\Delta_{11} + \Delta_{12} = 2\Delta_{12}$. With this choice of death rates, we see that the epistatic parameter E is zero, Now, (II.6) becomes

$$dp_0/d\sigma = \Delta p_0(1 - p_0). \tag{II.7}$$

This is the logistic equation which is easily solved. The solution is indicated in Fig. II.1. Note that, as expected, the A gene slowly dominates the population since $p_0(\sigma) \to 1$ as $\sigma \to \infty$.

(b) *Homozygosis*. $0 = \Delta_{11} = \Delta_{12} < \Delta = \Delta_{22}$. Now, $E < 0$ and again we should expect the A gene to dominate, but the approach to dominance should be slower than in the previous case. In fact, (II.6) becomes

$$dp_0/d\sigma = \Delta p_0(1 - p_0)^2,$$

whose solution approaches one as $\sigma \to \infty$, but now $1 - p_0$ is proportional to $1/\sigma$ whereas this difference was exponential in the previous case. The solution is indicated in Fig. II 1.

(c) *Heterozygosis*: $0 = \Delta_{12} < \Delta_{11}, \Delta_{22}$. In this case the heterozygotes have a selective advantage over either homozygote This is a case of practical interest, since it describes a mechanism for a population to maintain a polymorphism, i.e., heterozygotes, which are observed in many populations. With this choice, (II.6) becomes

$$dp_0/d\sigma = \Delta p_0(1 - p_0)(l - p_0),$$

where $\Delta = 2E$ and $l = (\Delta_{22} - \Delta_{12})/\Delta$. It is easy to see that all solutions of this equation with $0 < p_0(0) < 1$ approach l as $\sigma \to \infty$. In fact, for $p_0(0) < l$, the solution has a positive derivative, and so is an increasing function of σ. It cannot cross the solution $p_0(\sigma) \equiv l$ since through any point this equation has a unique solution, and this constant value defines a solution. Thus, it is bounded above and so has a limit as $\sigma \to \infty$. From the equation, the only possible limits for solutions (i.e., values where $(dp_0/d\sigma)(\infty) = 0$) are $p_0 = 0, l, 1$. Therefore, if $p_0(0) \leqq l$, $p_0(\infty) = l$. A similar argument works if $p_0(0) > l$.

This shows that the genotype distributions approach the values

$$D_0(\infty) = l^2, \quad 2H_0(\infty) = 2l(1 - l), \quad R_0(\infty) = (1 - l)^2,$$

as $\sigma \to \infty$. The solution for p_0 is indicated in Fig. II.1.

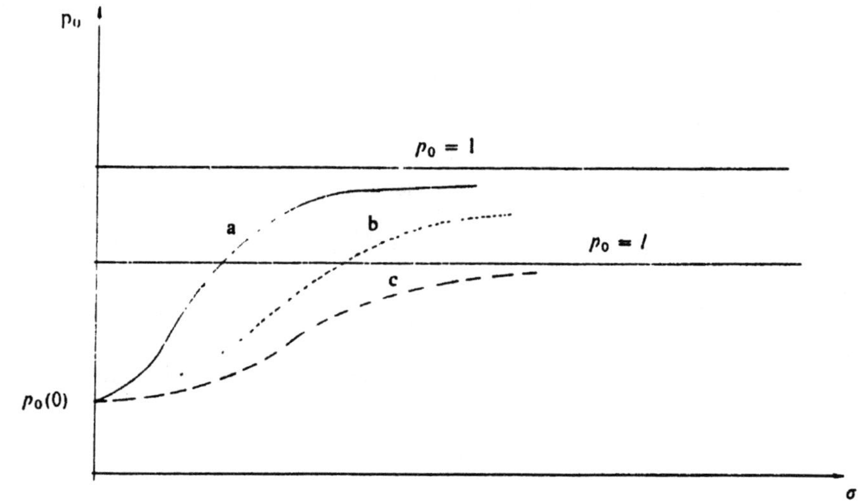

FIG II 1

2.3. Lethal genes. In this situation, we assume some of the genotypes are sterile. This is reflected in that their birth rates are zero. For simplicity, we shall also assume that $\Delta_{ij} = 0$ for all genotypes.

(a) *Lethal recessive genes*: $b_{11} = b_{12} = \beta > 0 = b_{22}$. With this choice of birth rates, the aa genotypes are sterile. In practice, this could happen if a is a recessive gene, thereby giving aa genotypes a distinctive phenotype, and these phenotypes are sterilized. This is frequently done in animal and plant breeding. In this model, the sterile individuals are allowed still to take part in the random mating. The case where they are removed from the mating population can be treated in a similar way by applying this analysis to the model for assortative mating which is derived in the next section.

The system (II.5) becomes

$$dD/d\sigma = \beta H^2,$$

$$2dH/d\sigma = \beta[(D + H) - (D^2 + 3HD + 4H^2)].$$

This system has two rest points: $D = 0$, $H = 0$ and $D = 1$, $H = 0$. In Fig. II.2, the phase plane diagram for this system is indicated. There are only two rest points; neither lies inside the triangle. Therefore, a standard phase plane argument shows that all solutions starting in the triangle, but with $D(0) \neq 0$ and $H(0) \neq 0$, must approach the rest point $D = 1, H = 0$. The arrows in Fig. II.2 show the direction a trajectory starting at various points on the triangle's boundary would proceed.

This analysis shows that the recessive lethal gene eventually dies out of the population. Of course, intuition predicts exactly that. However, the thing of

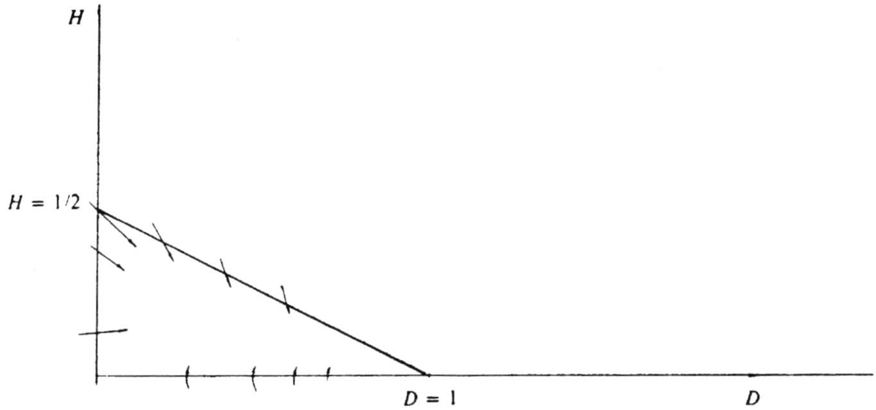

FIG II 2

interest here is the rate at which this occurs. Since $2H = 1 - D - R \leqq 1 - D$, we have

$$d(D - 1)/d\sigma \leqq \beta(D - 1)^2,$$

and so by integrating this inequality, we obtain

$$1 - D(\sigma) \geqq (1 - D(0))/[1 + (1 - D(0))\beta\sigma].$$

This estimate shows that the difference $1 - D(\sigma)$ approaches zero no faster than $1/\beta\sigma$ approaches zero!

(b) *Lethal dominant gene*: $b_{11} = b_{12} = 0 < \beta = b_{12}$. With this choice of birth rates, the system becomes

$$dD/d\sigma = -\beta D(1 - D - H),$$

$$2dH/d\sigma = \beta(D - H)(1 - D - H).$$

Now we see that $2p = 2D + 2H$ satisfies

$$2dp/d\sigma = -\beta(1 - p)p.$$

This is the logistic equation which can be easily solved to show that $p(\sigma)$ approaches zero at an exponential rate $\beta/2$ as $\sigma \to \infty$.

Thus, if we interpret $1/\beta$ as being the generation time for the population we see that in the case of a lethal dominant, the lethal gene is effectively out of the population after two generations, while the lethal recessive takes many generations to be removed to the same extent.

2.4. The De Finetti diagram. Geneticists display the genotype frequencies by plotting them simultaneously. This is easy to do because, although there are three genotypes, the relation $D + 2H + R = 1$ is satisfied. Therefore, the point $(D, 2H, R)$ in three dimensions always lies in a triangular region. De Finetti (1926) suggested the following representation: Take an equilateral triangle with altitude equal one

The vertices are labelled AA, Aa and aa as in Fig. II.3. The point $(D, 2H, R)$ is located in this triangle by drawing a line parallel to the side opposite AA at a distance D from it and a line parallel to the side opposite aa at a distance R from it. The intersection of these lines denotes the point $(D, 2H, R)$.

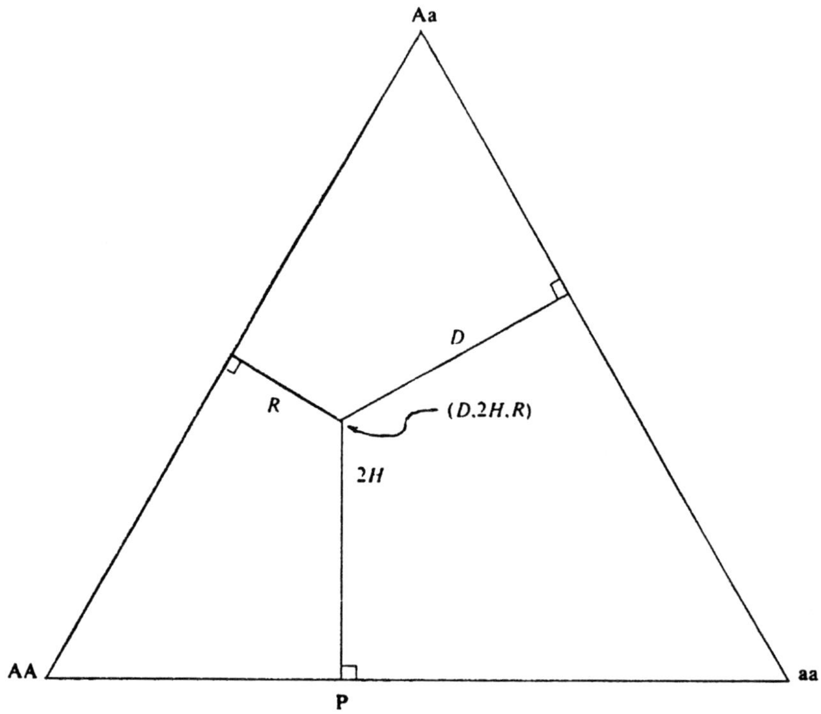

FIG. II.3

Recall that for an equilateral triangle, the sum of the perpendicular distance to the sides is equal to the altitude. Hence, $D + 2H + R = 1$. Moreover, the line segment from AA to P in Fig. II.3 is to the line segment from P to aa as $R + H$ is to $D + H$. This shows

$$\overline{\mathrm{AA\,P}} = 2q/\sqrt{3} \quad \text{and} \quad \overline{\mathrm{P\,aa}} = 2p/\sqrt{3}.$$

In Figure II.4, the locus of points which satisfy $D = p^2$, $H = pq$, $R = q^2$, as p ranges from zero to one, is the parabola in the diagram.

It was shown in § 2.2 that if there is no selection, the genotypes approach the Hardy-Weinberg proportions $(p^2, 2pq, q^2)$ which remain constant. Therefore, an initial point $(D(0), 2H(0), R(0))$ will approach a point on the curve C in Fig. II.4 as $t \to \infty$. Which point will it approach? Since the gene frequencies are constant, the point $(D(t), 2H(t), R(t))$ will move to C along the line through $(D(0), 2H(0), R(0))$ which is perpendicular to the face opposite Aa; that is, along the line $D + H = p$. This is illustrated in Fig. II.5.

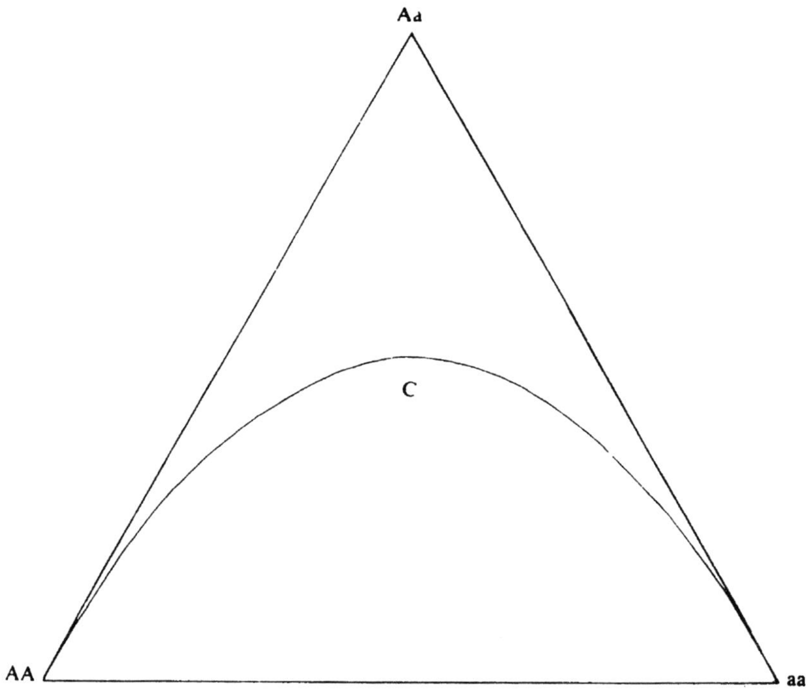

FIG. II.4

This schematic presentation also gives an interesting description of the phenomena of slow selection. For example, in the case of slow selection with the genotype AA favored, the solution $(D(t), 2H(t), R(t))$ quickly approaches Hardy-Weinberg proportions which then slowly change and approach $p = 1$. Figure II.6 illustrates this case. The curve in Fig. II.6 is the approximation to $(D(t), 2H(t), R(t))$ obtained in § 2.2.

2.5. Assortative mating and mutation. The basic model gives an adequate description of A gene frequencies in a great many situations. This is the case for large randomly mating populations where mutations which occur can be neglected. In many populations though, mating is not random for various reasons. For example, it may be that only certain phenotypes are allowed to mate, as in animal breeding. Also, there may be various other mechanisms which do not allow random mating such as social structure or geography. These sorts of situations can be accounted for by suitable modifications in the general model. This will be illustrated here for the case where a mating bias is introduced through phenotypes. Also, later in this section, the basic model will be modified to account for mutations. Incidentally, it is easy to account for migrations of genotypes into and out of the population. The modifications needed for this phenomenon are in principle the same as those needed for mutations, and will only be listed at the end of this section.

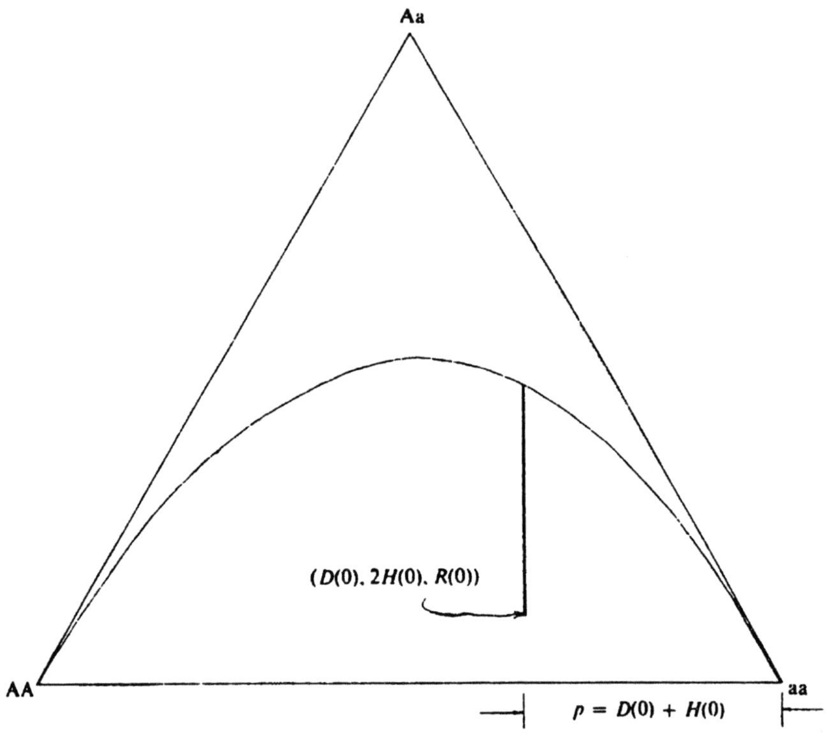

FIG. II 5

(a) *Assortative mating*. Suppose that A is the dominant gene and that a proportion α of the dominant phenotypes mate only with dominant phenotypes, and that a proportion β of the recessive phenotypes mate only with recessive phenotypes.

A dominant female is randomly mating in both the assortative and nonassortative pools of dominants The population can be divided into these groups in the following way:

$$n = n_{11} + n_{12} + n_{22} = \alpha(n_{11} + n_{12}) + (1 - \alpha)n_{11} + (1 - \alpha)n_{12} + \beta n_{22} + (1 - \beta)n_{22},$$

where the first term gives the number of dominant assorting phenotypes, etc. In fact, since the number of dominant phenotypes is $n_{11} + n_{12}$, the number assortatively mating is $\alpha(n_{11} + n_{12})$. Therefore, the proportion of dominant *phenotypes* (types AA or Aa) who are dominant *genotypes* (AA) is

$$\alpha D/(D + 2H) = \alpha n_{11}/(n_{11} + n_{12}).$$

Therefore, the number of progeny to dominant females due to fertilizations by dominant assortatively mating males is

$$b_{11}n_{11}\alpha D/(D + 2H).$$

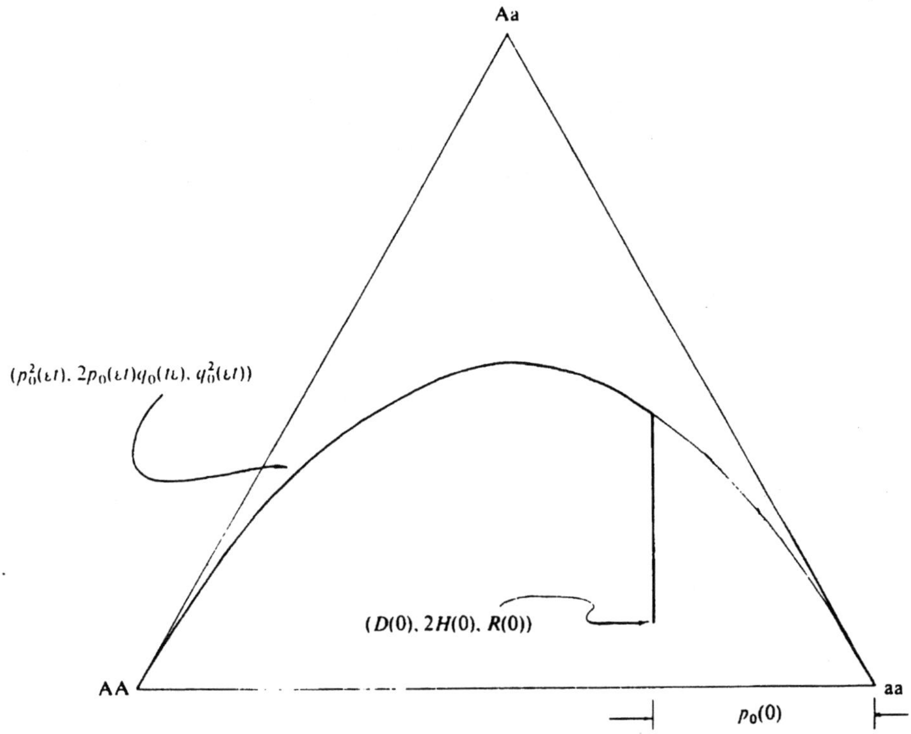

FIG II 6

Similar arguments lead to the equation

$$\frac{dn_{11}}{dt} = b_{11}n_{11}\left\{\frac{\alpha D}{D+2H} + (1-\alpha)D + \frac{\alpha H}{D+2H} + (1-\alpha)H\right\}$$
$$+ b_{12}n_{12}\left\{\frac{\alpha D}{2(D+2H)} + \frac{\alpha H}{2(D+2H)} + (1-\alpha)\frac{D}{2} + (1-\alpha)\frac{H}{2}\right\} - d_{11}n_{11}.$$

Similar equations can be derived for n_{12} and n_{22}. As in the basic model, these equations are used to compute $dD/dt = d(n_{11}/n)/dt$, etc. The results are

$$dD/dt = b_{11}D\left[\frac{\alpha p}{D+2H} + (1-\alpha)p\right] + b_{12}H\left[\frac{\alpha p}{D+2H} + (1-\alpha)p\right]$$
$$- d_{11}D - \bar{m}D,$$

$$2dH/dt = b_{11}D\left[\frac{\alpha H}{D+2H} + (1-\alpha)H + (1-\beta)R\right]$$
$$+ b_{12}H[\alpha + (1-\alpha)(D+2H) + (1-\beta)R]$$
$$+ b_{22}R[(1-\alpha)D + (1-\alpha)H] - 2d_{12}H - 2\bar{m}H,$$

$$dR/dt = b_{12}H\left[(1-\alpha)H + \frac{\alpha H}{D+2H} + (1-\beta)R\right]$$
$$+ b_{22}R[(1-\alpha)H + R] - d_{22}R - \bar{m}R.$$

It is easy to check that when $\alpha = \beta = 0$, the basic model is again obtained.

The case where the recessive phenotypes are sterilized was mentioned in the last section. If these are not allowed to mate with the dominant phenotypes, then we must take $\alpha = \beta = 1$, $b_{22} = 0$, $b_{11} = b_{12} = \beta_0 > 0$, and $d_{11} = d_{12} = d_{22} = 0$ to obtain the correct mating pattern On taking $\alpha = \beta = 1$, $b_{22} = 0$, $b_{11} = b_{12} = \beta_0 > 0$, and $d_{11} = d_{12} = d_{22} = 0$, we have

$$dD/dt = \beta_0 \frac{p^2}{D + 2H} - \beta_0(D + 2H)D,$$

$$2dH/dt = 2\beta_0 Hp/[D + 2H] - 2\beta_0(D + 2H)H,$$

$$dR/dt = \beta_0 H^2/[D + 2H] - \beta_0(D + 2H)R.$$

It is easily shown that the solution of this system is qualitatively the same as for the lethal recessive.

(b) *Mutation.* Suppose the gene A can become an a gene and vice versa. Further, suppose that these changes occur at specified rates. μ and ν, respectively. Then in a time interval of length h, a number

$$\nu n_{12} h$$

of heterozygotes will become AA and a similar number aa. Also,

$$2\mu n_{11} h$$

aa's will become heterozygotes. The mutation model then can be constructed just as before, but with new terms added

$$dn_{11}/dt = \nu n_{12} - 2\mu n_{11} + \cdots,$$

$$dn_{12}/dt = 2\nu n_{22} + 2\mu n_{11} - \mu n_{12} - \nu n_{12} + \cdots,$$

$$dn_{22}/dt = \mu n_{12} - 2\nu n_{22} + \cdots,$$

where the dots indicate terms already present in the basic model. As a result, the gene frequency equation takes the form

$$dp/d\sigma = \nu(1 - p) - \mu p + \cdots,$$

and the equations for the genotype frequencies are modified in a similar way.

(c) *Migration.* To account for migration of genotypes, we introduce the migration rates and then modify the equations as in (b). For example, if there is an immigration of AA's at a rate k_{11}, then we would have

$$dn_{11}/dt = k_{11} + \cdots$$

The remaining modifications are obvious.

3. Age dependent population genetics. Mendelian genetics can be studied in age dependent populations by combining the principles of this chapter with the approach to age dependent populations described in Part I This could lead to a quite complete, but very complicated model, so only a sample problem will be described here.

The diploid population under consideration consists of males and females, which we assume are identically distributed with respect to genotypes determined by one locus having two alleles. Moreover, we suppose that the birth and death rates are independent of sex and depend only on age (not on time). Therefore, the male and female populations do not differ significantly.

Because of this, everything will be based on the female population. As in Part I, functions are introduced to describe the age distributions of both the female population and the population of couples. Let

$f(a, t)$ = age distribution of females at time t,

$c(a, a', t)$ = distribution of couples having male of age a and female of age a' at time t.

Then f and c satisfy the dynamic equations

$$\partial c/\partial a + \partial c/\partial a' + \partial c/\partial t = -(\Delta_1 + \Delta_2 + d)c + \mu S,$$

$$\partial f/\partial t + \partial f/\partial a = -\Delta_2 f,$$

where Δ_1, Δ_2 and d are the death rates of couples due to male death, female death and divorce, respectively. The boundary conditions for c are

$$c(a, 0, t) = c(0, a', t) = 0,$$

for all a, a', t. Finally, suppose there are sufficiently many males so that each widow or divorced female is mated again during the child-bearing ages. Therefore, we may take $\Delta_1 = d = 0$ in the equation for c. With these conditions, it is reasonable to assume that the marriage function is given by the number of eligible females.

$$S = f_e(a, t)$$

This amounts to taking ρ in Part I, §4 as

$$\rho(x) = \begin{cases} 1 + \dfrac{1}{x} & \text{for} \quad x > 1, \\ 1 + x & \text{for} \quad x < 1. \end{cases}$$

To simplify the computations in this section, we shall further assume that only males and females of the same age can mate: Therefore, we take $\mu(a, a', t) = \mu_0(a) \times \delta(a - a')$. Substituting all of these simplifications into the coupling equation, we have

$$\partial c/\partial a + \partial c/\partial a' + \partial c/\partial t = -\Delta_1 c + \mu_0(a)\delta(a - a')f_e(a', t).$$

It follows that if $c(a, a', 0) = \delta(a - a')f_0(a)$, then

$$c(a, a', t) = \delta(a - a')g(a)f(a, t),$$

where $g(a) = 1 - \exp[-\int_0^a \mu_0(a')\,da']$.

Since the goal is to study the dynamics of genotypes determined by one locus having two alleles, say **B** and **b**, functions are introduced to describe the age structure of each of these populations:

$$U(a,t) = \text{age distribution of BB at time } t,$$
$$2V(a,t) = \text{age distribution of Bb at time } t,$$
$$W(a,t) = \text{age distribution of bb at time } t.$$

These functions describe the numbers of genotypes. Each class is assumed to have its own birth and death schedules, which are denoted by

$$\lambda_U(a),\quad \lambda_V(a),\quad \lambda_W(a),\quad \text{and}\quad d_U(a),\quad d_V(a),\quad d_W(a).$$

Using the approach described in Part I and the identity

$$f(a,t) = U(a,t) + 2V(a,t) + W(a,t),$$

we derive the dynamic equations:

$$\begin{aligned} \partial U/\partial t + \partial U/\partial a &= -d_U(a)U, \\ \partial V/\partial t + \partial V/\partial a &= -d_V(a)V, \\ \partial W/\partial t + \partial W/\partial a &= -d_W(a)W, \\ \partial f/\partial t + \partial f/\partial a &= -(d_U U + 2d_V V + d_W W). \end{aligned} \tag{II.8}$$

The birth rate of females is clearly

$$f(0,t) = \int_0^\infty g(a)[\lambda_U(a)U(a,t) + 2\lambda_V(a)V(a,t) + \lambda_W(a)W(a,t)]\,da, \tag{II.9}$$

but the birth rates of the genotypes are somewhat more involved. Using Mendel's laws, we see that

$$\begin{aligned} U(0,t) &= \int_0^\infty g(a)\{\lambda_U U(U+V)/f + \lambda_V V(U+V)/f\}\,da, \\ 2V(0,t) &= \int_0^\infty g(a)[\lambda_U U(V+W)/f + \lambda_V V + \lambda_W W(U+V)/f]\,da, \\ W(0,t) &= \int_0^\infty g(a)[\lambda_V V(V+W)/f + \lambda_W W(W+V)/f]\,da. \end{aligned} \tag{II.10}$$

Finally, we set the initial conditions

$$\begin{aligned} &U(a,0) = U_0(a),\quad V(a,0) = V_0(a),\quad W(a,0) = W_0(a), \\ &f(a,0) = U_0(a) + 2V_0(a) + W_0(a), \end{aligned} \tag{II.11}$$

and we obtain from (II.8-11) a well-posed problem: If the fertilities $\lambda_U, \lambda_V, \lambda_W$, and the initial data are smooth, nonnegative and have bounded support and if d_U, d_V, d_W are smooth functions, then this system of equations has a unique nonnegative solution.

In 1926, Norton analyzed this system under assumptions that selection occurred only through death ($\lambda_U = \lambda_V = \lambda_W$) and that the births are in Hardy-Weinberg proportions ($V^2(0,t) = U(0,t)W(0,t)$). Rather than a description of that analysis, the no-selection case will be studied without the assumption of Hardy-Weinberg proportions.

Suppose $\lambda \equiv \lambda_U = \lambda_V = \lambda_W$ and $d \equiv d_U = d_W = d_V$ Then by defining

$$P(a,t) = U(a,t) + V(a,t),$$

we see that

$$P(0,t) = \int_0^\infty g(a)\lambda(a)P(a,t)\,da.$$

Also,

$$\partial P/\partial t + \partial P/\partial a = -dP, \quad P(a,0) = U_0(a) + V_0(a).$$

It was shown in Part I that the characteristic equation

$$1 = \int_0^\infty g(a)\lambda(a)\exp\left[-\int_0^a d(a')\,da'\right]e^{-qa}\,da$$

has a unique real solution for q. Moreover, for that choice of q, the limit

$$\lim_{t\to\infty} e^{-qt}P(a,t),$$

exists and defines a smooth "stable age distribution" $P^*(a)$. $P^*(a)$ satisfies the equation

$$dP^*/da = -[d(a) + q]P^*$$

and $P^*(0)$ can be evaluated explicitly by Laplace transform methods. Since $f(a,t)$ satisfies the same problem but with larger initial data, we have that

$$\lim_{t\to\infty} e^{-qt}f(a,t)$$

exists and defines a smooth function $F(a)$. Since $F(0) > 0$, it follows that for all large t, $f(a,t) \neq 0$ for $a \leqq t$.

Now let us return to the problems for U, V and W. First, we have that

$$e^{-qt}U(0,t) = \int_0^\infty g(a)\lambda(a)[e^{-2qt}P^2(a,t)/e^{-qt}f(a,t)]\,da.$$

Therefore, this limit (as $t \to \infty$) exists and defines

$$\lim_{t\to\infty} e^{-qt}U(0,t) = P_0^{*2}(0)/F(0).$$

Similarly, we can show that

$$\lim_{t\to\infty} e^{-qt}V(0,t) = P^*(0)Q^*(0)/F(0) \quad \text{and} \quad \lim_{t\to\infty} e^{-qt}W(0,t) = Q^{*2}(0)/F(0),$$

where $Q^*(0) = F(0) - P^*(0)$. Let $U^*(a)$ be the stable solution of

$$dU^*/da = -(d + q)U^*, \quad U^*(0) = P^{*2}(0)/F(0).$$

Then the difference $\Delta = e^{-qt}U(a,t) - U^*(a)$ satisfies

$$\frac{\partial \Delta}{\partial t} + \frac{\partial \Delta}{\partial a} = -(d+q)\Delta, \quad \Delta(0,t) = e^{-qt}U(0,t) - \frac{P^{*2}(0)}{F(0)},$$

and $\Delta(a,0) = U_0(a) - U^*(a)$. Since $\Delta(0,t) \to 0$ as $t \to \infty$, it is easy to see that $\Delta(a,t) \to 0$ as $t \to \infty$ for all a

In the same way it can be shown that

$$\lim_{t \to \infty} e^{-qt} 2V(a,t) = \frac{2P^*(0)Q^*(0)}{F(0)} e^{-qa - \int_0^a d(a')da'} \equiv 2V^*(a),$$

and

$$\lim_{t \to \infty} e^{-qt} W(a,t) = \frac{Q^{*2}(0)}{F(0)} e^{-qa - \int_0^a d(a')da'} \equiv W^*(a).$$

Therefore, all the genotypes approach stable age distributions.

Finally, notice that

$$U(a,t)/f(a,t) = e^{-qt}U(a,t)/e^{-qt}f(a,t) \to U^*(a)/F(a) = P^{*2}(0)/F^2(0).$$

Therefore, the frequency of the genotypes approaches a Hardy-Weinberg distribution. In fact,

$$(U(a,t)/f(a,t), 2V(a,t)/f(a,t), W(a,t)/f(a,t))$$
$$\to (P^{*2}(0)/F^2(0), 2P^*(0)Q^*(0)/F^2(0), Q^{*2}(0)/F^2(0))$$

as $t \to \infty$, which is independent of age!

4. Propagation of a gene in a spatially distributed population. In Part III models for the spread of infection in a geographically distributed population will be extensively studied. Similar models can be constructed to describe the propagation of genes in a spatially distributed population. An example of such a model is described and analyzed in this section.

Rather than repeat much of the earlier discussion to describe in detail how the model is derived, let us simply consider a population's gene pool consisting of two alleles, A and a. The population is viewed as being spread out on a straight line, and at a point x on the line, the frequency of the allele A in the gene pool is given by $p(x,t)$. Suppose that at the point x, the genotypes have fitnesses given in the following table,

genotypes	AA	Aa	aa
fitness	$m_{11}(x)$	$m_{12}(x)$	$m_{22}(x)$,

where $m_{11} = b_{11} - d_{11}$, etc. Suppose that there is no dominance: $m_{11} + m_{22} = 2m_{12}$. Then we take for p the equation

$$\partial p/\partial t = \kappa(\partial^2 p/\partial x^2) + s(x)p(1-p), \tag{II.12}$$

where κ measures the diffusion of the gene along the line and $s(x)$ denotes the selective coefficient

$$s(x) = m_{11}(x) - m_{12}(x)$$

The term

$$s(x)p(1 - p),$$

is obtained just as it was in the no epistasis cases in §2.2(a). In fact, if $\kappa = 0$, this equation reduces to exactly the "slow time" equation for the A-gene frequency at the point x, when there is no epistasis. This remark indicates how this equation is derived. The diffusion term $\kappa\partial^2 p/\partial x^2$ arises as shown in the diffusion approximation in Part III, § 2.3.

$s(x)$ will be constant in the following sections. The case where it is nonconstant is studied by Fleming (1974).

The equation (II.12) was proposed by Fisher (1936), and it will be referred to as Fisher's equation here.

Fisher's equation for the propagation of a favored gene. Suppose now that s is a positive constant. Therefore, A is the allele favored by selection at each point, and it is expected eventually to dominate the gene pool at each point where it is present. To see how this occurs, let us suppose the gene pools are described at $t = 0$ by

$$\text{(II.13)} \qquad \begin{aligned} p(x, 0) &= 1 \quad \text{for} \quad x < 0, \\ &= 0 \quad \text{for} \quad x > 0 \end{aligned}$$

It is expected that the A-gene will diffuse to the right of $x = 0$, and eventually establish itself at each point $x > 0$. Indeed, this is what happens.

Consider the equations (II.12) along with the boundary conditions

$$\text{(II.14)} \qquad p(-\infty, t) = 1, \qquad p(\infty, t) = 0, \qquad 0 \leqq p \leqq 1.$$

The problem (II.12, 14) was studied by Kolmogoroff, Petrovsky and Piscounov (1936). Their approach was to show that there are progressing wave solutions and that the initial data (II.13) evolves into one of these progressing waves.

To find the progressing waves, we seek a solution of (II.12, 14) in the form

$$p = p(\xi), \qquad \xi = x - ct.$$

The equation for this function then is

$$\text{(II.15)} \qquad -c(dp/d\xi) = (d^2p/d\xi^2) + sp(1 - p).$$

An argument like the phase plane argument used in Part III, § 3, establishes the following result.

THEOREM (Kolmogoroff et al.). *Let s and κ be positive constants. Then for each $c \geqq c_{\min} = 2\sqrt{\kappa s}$, there is a unique solution of* (II.15), *$p_c(\xi)$ which satisfies $0 \leqq p \leqq 1$, $p(-\infty) = 1$, $p(\infty) = 0$. Moreover, for $c < c_{\min}$ there is no such solution.*

4.1. The form of the progressing waves. The progressing waves can be analyzed through a perturbation scheme like the one used earlier. A dimensionless parameter is introduced by

$$\varepsilon = c_{\min}^2/c^2.$$

Substituting this in (II.15) along with the new variables

$$\zeta = \sqrt{\varepsilon}\,\xi, \qquad p(\xi) = v(\zeta),$$

we obtain

$$\varepsilon\kappa(d^2v/d\zeta^2) + \sqrt{s\kappa}\,(dv/d\zeta) + sv(1 - v) = 0. \tag{II.16}$$

The solution of this equation which satisfies the boundary conditions $v(-\infty) = 1$, $v(\infty) = 0$, $0 \leqq v \leqq 1$, is to be found. This problem can be converted to two initial value problems, one for $\zeta \geqq 0$ and one for $\zeta \leqq 0$, by finding the inflection point of the progressing wave. We do this next.

Equation (II 16) can be converted to the system

$$dv/d\zeta = w, \qquad \varepsilon\kappa\, dw/d\zeta = sv(v - 1) - \sqrt{s\kappa}\, w.$$

The inflection point occurs when $w' = v'' = 0$, and the following analysis determines its location.

The first order system is equivalent to the equation

$$\varepsilon\, dw/dv = [sv(v - 1) - \sqrt{s\kappa}\, w]/\kappa w. \tag{II.17}$$

If we look for the solution of (II.17) in the form

$$w = w_0(v) + \varepsilon w_1(v) + \cdots,$$

then we find

$$w_0(v) = sv(v - 1)/\sqrt{s\kappa},$$
$$w_1(v) = -\kappa w_0[dw_0/dv]/\sqrt{s\kappa}, \quad \text{etc.}$$

The result of this formal calculation is that

$$w(v) = [sv(v - 1)/\sqrt{s\kappa}] - \varepsilon[w_0 w_0'/\sqrt{s\kappa}] + 2\varepsilon^2[w_1 w_0' + w_0 w_1']/\sqrt{s\kappa} + \varepsilon^3\{\cdots\}.$$

In the terminology of perturbation theory, we have determined an expansion for the outer solution of the problem.

This expansion is used to determine the inflection point. It occurs when $d^2v/d\zeta^2 = 0$, which happens when $w'(v) = 0$. Now,

$$w'(v) = s[2v - 1]/\sqrt{s\kappa} - \varepsilon(w_0 w_0'' + (w_0')^2)/\sqrt{s\kappa}$$
$$+ 2\varepsilon^2[w_1' w_0' + 2w_1 w_0'' + w_0 w_1'']/\sqrt{s\kappa} + \varepsilon^3\{\cdots\}.$$

Now, let us determine v as a series in ε so that $w'[v(\varepsilon)] = 0$.

$$v = v_0 + v_1\varepsilon + v_2\varepsilon^2 + \cdots.$$

The coefficients in this expansion can be determined successively from the expansion for w':

$$\sqrt{\frac{s}{\kappa}}[2v_0 - 1] = 0, \quad \sqrt{\frac{s}{\kappa}}[2v_1] = \frac{1}{\sqrt{s\kappa}}[w_0(v_0)w_0''(v_0) + (w_0'(v_0))^2], \quad \text{etc.}$$

Thus,

$$v_0 = 1/2, \quad v_1 = (1/s)[w_0(1/2)w_0''(1/2) + (w_0'(1/2))^2], \quad \text{etc}$$

The inflection points of the progressing waves determined by the theorem for small ε (i.e., $c \gg c_{\min}$) occur at

$$v = (1/2) + (\varepsilon/s)[w_0(1/2)w_0''(1/2) + (w_0'(1/2))^2] + \varepsilon^2\{\cdots\},$$
$$w = -\sqrt{s/16\kappa} + \varepsilon^2\{\cdots\}.$$

Therefore, we may replace the boundary value problem (II.16) by the initial value problem

$$\varepsilon\kappa\frac{d^2v}{d\zeta^2} + \sqrt{s\kappa}\frac{dv}{d\zeta} + sv(1 - v) = 0,$$
$$v(0) = (1/2) + \varepsilon v_1 + \varepsilon^2\{\cdots\},$$
$$(dv/d\zeta)(0) = -(1/4)\sqrt{s/\kappa} + \varepsilon^2\{\cdots\}.$$

The method of matched asymptotic expansions can be applied to show that this problem has a unique solution on $-\infty < \zeta < \infty$, and it has a uniformly valid approximation

$$v(\zeta) = V_0(\zeta) + \varepsilon V_1(\zeta) + \varepsilon^2\{\cdots\}.$$

The coefficients are determined as follows: V_0 satisfies

$$\sqrt{s\kappa}\,(dV_0/d\zeta) + sV_0(1 - V_0) = 0, \quad V_0(0) = 1/2,$$

and so

$$V_0(\zeta) = 1/(1 + \exp\zeta).$$

A similar problem shows that

$$V_1(\zeta) = \frac{e^\zeta}{(1 + e^\zeta)^2}[4V_1(0) + \log(\zeta e^\zeta/(1 + e^\zeta)^2)], \quad \text{etc.}$$

This analysis leads us to expect that the progressing wave given by the theorem will be approximated by

$$V_0\left[\frac{2\sqrt{s\kappa}}{c^2}(x - ct)\right] + \frac{2\sqrt{s\kappa}}{c^2}V_1\left[\frac{2\sqrt{s\kappa}}{c^2}(x - ct)\right],$$

at least for $c \gg c_{\min} = 2\sqrt{s\kappa}$ (i.e., $\varepsilon \ll 1$).

This is essentially the analysis carried out by Canosa (1972). He compared this approximation with numerically computed solutions. The results agreed to three decimal places, even for $\varepsilon = 1$!

This analysis gives, with very little effort, a great deal of insight into the analytic structure of the progressing wave solutions.

4.2. Stability of progressing waves: $c = c_{\min}$. Fisher gave an heuristic argument to show that the solution of (II.12) which has the special initial state given by (II.13) will evolve into a wave progressing with velocity $c_{\min}$. The analysis which is briefly described here is a translation of Fisher's argument into the context of a diffusion process. Only the method and results are stated in this section. The details are presented for a more general diffusion process in the context of epidemic spread (see Part III, § 2.3(c)).

The total A-gene population which lies to the right of some point R is given by the integral

$$\int_R^\infty p(x,t)\,dx$$

For short times and large R, this population will be very small since p was initially zero. Let us attempt to determine how R must change with time so that this population remains constant. To do this, fix a population size, say P, and let $R(t)$ be defined by the equation

$$\int_{R(t)}^\infty p(x,t)\,dx = P.$$

If P is small, p^2 is very small and the equation (II.12) can be written approximately as

$$\frac{\partial p}{\partial t} = \kappa\frac{\partial^2 p}{\partial x^2} + sp.$$

As shown in Part III, § 2.3(c), this equation can be solved explicitly in terms of the fundamental solution of the heat equation It follows that

$$\int_R^\infty p(x,t)\,dx \sim 2\exp[-(R^2/4\kappa t) + st] \quad \text{for} \quad R \gg 1.$$

Therefore, $2\exp[-(R^2/4\kappa t) + st] = P$, and taking logarithms, we obtain

$$R^2(t) = 4\kappa st^2 + t\log(P/2).$$

In particular, for large t, the velocity of $R(t)$, dR/dt, is approximately

$$dR/dt = 2\sqrt{\kappa s} \equiv c_{\min}.$$

This argument provides some evidence that the initial data (II.13) evolve into a wave progressing with velocity $c_{\min}$. An actual proof of this fact was proposed by Kolmogoroff et al

4.3. Stability of progressing waves: $c > c_{\min}$. The progressing wave solutions given in the theorem exhibit curious stability properties. We have seen that initial data like that in (II.13) (i.e., $p(x,0) = H(-x)$) evolve into progressing waves which propagate with the velocity $c_{\min}$. Waves having velocities greater than $c_{\min}$ are stable to certain kinds of perturbations. It will be shown that the rate at which initial data approach zero and one as x approaches ∞ and $-\infty$, respectively, determines the velocity of the progressing wave that it will evolve into. The slower the approach, the higher velocity the progressing wave will have. Thus, the long time behavior depends critically and very sensitively on the behavior of the initial data at $x = \pm\infty$.

The stability of travelling waves for $c > c_{\min}$ can be investigated in the following way. Suppose that $u_c(x - ct)$ is a travelling wave solution of (II.12) for some $c > c_{\min}$, take $\kappa = 1$ and s to be a constant. Restrictions on initial data $p(x, 0)$ will be determined which ensure that

$$\sup_x |p(x, t) - u_c(x - ct)| \to 0$$

as $t \to \infty$. Let

$$\tilde{\psi}(x, t) = p(x, t) - u_c(x - ct).$$

This function satisfies

$$\frac{\partial \tilde{\psi}}{\partial t} = \frac{\partial^2 \tilde{\psi}}{\partial x^2} + s(\tilde{\psi} + u_c)(1 - \tilde{\psi} - u_c) - su_c(1 - u_c).$$

Setting $z = x - ct$ and $\psi(z, t) = \tilde{\psi}(x, t)$, we obtain

$$\text{(II.18)} \qquad \frac{\partial \psi}{\partial t} = \frac{\partial^2 \psi}{\partial z^2} + c\frac{\partial \psi}{\partial z} + s(1 - 2u_c(z))\psi + \mathscr{H}(\psi),$$

where $\mathscr{H}(\psi) = o(|\psi|)$ as $|\psi| \to 0$. The first step is estimating the solution of the linear part of this equation. To begin, we have the following lemma.

LEMMA. *For $c > 2\sqrt{s}$, there exists a constant $\mu = \mu_c > 0$ and a function $\varphi = \varphi_c \in C^2(-\infty, \infty)$ such that*

$$(d^2\varphi/dz^2) + c(d\varphi/dz) + s(1 - 2u_c(z))\varphi = -\mu\varphi,$$

$\varphi(z) > 0$ and $\varphi(\pm\infty) = 0$.

The proof of this lemma follows from a straightforward phase plane argument, when polar coordinates are used.

Now, consider the linear problem

$$\text{(II.19)} \qquad \frac{\partial \hat{\psi}}{\partial t} = \frac{\partial^2 \hat{\psi}}{\partial z^2} + c\frac{\partial \hat{\psi}}{\partial z} + s(1 - 2u_c(z))\hat{\psi}.$$

The change of variables

$$\hat{\psi} = e^{-\mu t}\phi(z)\Psi,$$

takes this problem into the problem for Ψ:

$$\frac{\partial \Psi}{\partial t} = \frac{\partial^2 \Psi}{\partial z^2} + \left(\frac{2\phi_z}{\phi} + c\right)\frac{\partial \Psi}{\partial z}.$$

The maximum principle can be applied to this equation to show that

$$|\Psi(z, t)| \leqq \sup_{\hat{z}} |\Psi(\hat{z}, 0)|,$$

provided the right side is bounded. Therefore,

$$|\hat{\psi}(z, t)| \leqq \phi(z)\, e^{-\mu t} \sup_{\hat{z}} |\hat{\psi}(\hat{z}, 0)/\phi(\hat{z})|,$$

if the initial condition $\hat{\psi}(z, 0)$ satisfies

$$\hat{\psi}(z, 0) = O(\varphi(z)) \quad \text{as} \quad z \to \pm\infty$$

These estimates are interpreted as estimates of the fundamental solution applied to $\hat{\psi}(z, 0)$ in the following way. Let the fundamental solution of (II.19) be denoted by $\Gamma(z, \zeta; t)$. Then the above estimate shows that

$$\left| \int_{-\infty}^{\infty} \Gamma(z, \zeta; t)\hat{\psi}(\zeta, 0)\, d\zeta \right| \leqq \phi(z)\, e^{-\mu t} \sup_{\hat{z}} |\hat{\psi}(\hat{z}, 0)/\phi(\hat{z})|.$$

Also, using Γ, we can rewrite the problem (II.18) as an equivalent integral equation

$$\psi(z, t) = \int_{-\infty}^{\infty} \Gamma(z, \zeta; t)\psi_0(\zeta)\, d\zeta + \int_0^t \int_{-\infty}^{\infty} \Gamma(z, \zeta, t - t')\mathscr{H}(\psi(\zeta, t'))\, d\zeta\, dt'.$$

Taking norms and using the estimate of Γ, we have

$$|\psi(z, t)| \leqq \phi(z)\, e^{-\mu t} \sup_{\zeta} |\psi_0(\zeta)/\phi(\zeta)| + \int_0^t \phi(z)\, e^{-\mu(t - t')} \sup_{\hat{z}} \left| \frac{\mathscr{H}(\psi(\hat{z}, t'))}{\phi(\hat{z})} \right| dt'.$$

Since $\mathscr{H} = o(|\psi|)$, a standard argument using Gronwall's inequality shows that there is $\rho > 0$ such that

$$|\psi(z, t)| \leqq \varphi(z)\, e^{-\mu t/2} \sup_{\hat{z}} |\psi_0(\hat{z})/\varphi(\hat{z})|.$$

provided $|\psi_0(z)/\varphi(z)| < \rho$. This inequality proves the following result.

THEOREM. *Let $c > 2\sqrt{s}$ and let μ_c and φ_c be as in the lemma. Then there is $\rho > 0$ such that for any function $\psi_0(z)$ which satisfies*

$$\sup |\psi_0(z)/\varphi_c(z)| \leqq \rho,$$

the solution of (II.12) *satisfying*

$$p(x, 0) = \psi_0(x) + u_c(x)$$

can be written as

$$p(x, t) = u_c(x - ct) + \psi(x - ct, t),$$

where u_c is a travelling wave having velocity c and ψ satisfies

$$\sup_x |\psi(x - ct, t)| \leqq K\, e^{-\mu t/2},$$

for some constant $K > 0$.

The condition that $\psi_0(z) = O(\varphi_c(z))$ as $z \to \pm\infty$, guarantees that if $\psi_0(z)$ has the correct asymptotic behavior as $z \to \pm\infty$ (determined by c and μ), then the solution it defines evolves into a wave with speed of propagation c. In particular, it shows that the domains of attraction of the various travelling waves are determined by asymptotic conditions at $z = \pm\infty$, where p is near zero or one. This is a strongly negative result. In fact, the criteria for being in a particular domain could never be verified for real data. More important is the fact that membership in a domain depends on an initial function's behavior near $p = 0$ and $p = 1$, precisely the regions where the random effects, which have not been accounted for in the model, are dominant.

BIBLIOGRAPHY

D BONNER AND S MILLS (1964) *Heredity*. 2nd ed., Prentice-Hall, Englewood Cliffs, N J.

J CANOSA (1973) *On a nonlinear equation of evolution*, IBM J Res and Dev, 17, pp 307–313

L L CAVALLI-SFORZA AND W F BODMER (1971) *The Genetics of Human Populations*, W H Freeman, San Francisco

J F CROW AND M KIMURA (1970) *An Introduction to Population Genetics Theory*, Harper-Row, New York

B DE FINETTI (1926) *Considerazioni matematiche sur l'ereditarieta Mendeliana*. Metron, 6, pp 1–41.

R A FISHER (1936) *The wave of advance of advantageous genes*, Ann Eugen, 7, pp 355–369

——— (1958) *The Genetical Theory of Natural Selection*, 2nd ed, Dover, New York.

W. FLEMING (1974) *A nonlinear parabolic partial differential equation in population genetics*, Proc. Int. Conf on Diff Eqns, University of Southern California, Los Angeles.

J B S HALDANE (1932) *The Causes of Evolution*, Harper-Row, New York

H HARRIS (1966) *Human Biochemical Genetics*, Cambridge University Press, London

F HOPPENSTEADT (1971) *Properties of solutions of ordinary differential equations with small parameters*, Comm Pure Appl Math, 24, pp 807–840

I M LERNER (1968) *Heredity, Evolution and Society*, W H Freeman, San Francisco

V A. MCKUSICK (1964) *Human Genetics*. Prentice-Hall, Englewood Cliffs, N J

P A. P. MORAN (1962) *The Statistical Processes of Evolutionary Theory*, Clarendon Press, Oxford.

H T J NORTON (1926) *Natural selection and Mendelian variation*, Proc London Math Soc. Ser. 2, 28, pp 1–45

A. KOLMOGOROFF, I PETROVSKY AND N PISCOUNOV (1936) *Etude de l'equation de la diffusion avec croissance de la quantite de matière et son application à un problème biologique*. Moscow University, Bull Mathematique, Ser Internationale Ser A 1, no 6, pp. 1–25

B. WALLACE (1968) *Topics in Population Genetics*, W W. Norton, New York

S. WRIGHT (1968, 1969) *Evolution and the Genetics of Populations Vol I Genetics and Biometric Foundations, Vol II The Theory of Gene Frequencies*, University of Chicago Press, Chicago

III. Theories of Epidemics

1. General theory of contagious phenomena. This part begins with several models of epidemics which have been extensively studied. Next, a description and analysis of a general age dependent mathematical theory for contagious phenomena is carried out.

1.1. Some examples of epidemic theories. The examples describe a population which is partitioned into several distinct classes by an infection. In particular, the classes are the susceptibles (denoted by S), exposed but not yet infectious (E), infectives (I), quarantined infectives (Q), and individuals removed from the process (R). In actual epidemic situations there usually would not be just one kind of susceptible, infective, etc.; moreover, age dependent structure of the population as well as the various classes frequently play important roles. While models for some of these more complicated features will be formulated, attention will be focused on very simple models.

This will illustrate one of the main reasons for studying epidemic models: Quite complicated systems often have features which are correctly described by much simpler versions. When this is the case, and these particular features are the things of interest, it is necessary only to consider a simple model. It is a primary problem to determine when this situation occurs. The main result of this kind to be described here is about threshold phenomena where it is shown that the addition of a complicated population structure does not dramatically change that particular qualitative feature. This analysis is carried out in § 2.

It is convenient to use a compartmental description for the various models. In this, the symbols denoting the various classes to be considered are written and arrows are drawn which indicate the allowable transitions between classes. When the rates for certain transitions are specified, these are written over the particular arrow. On the other hand, if individuals remain in a class for a fixed holding time, this is indicated by a superscript For example, the diagram

$$\xrightarrow{A} S \xrightarrow{rIS} I^{\sigma} \rightarrow$$

involves only the susceptible and infective classes. It indicates that susceptibles are gained at a given rate A and lost at a rate proportional to the number of susceptibles and infectives. Then infectives are gained at the same rate. Finally, after remaining in the infective class for a fixed length of time σ, individuals leave I.

In order to study the dynamics of such a system, time dependent functions are introduced to measure the content of the various classes. For example, in the cases above two functions would be introduced, $S(t)$ to measure the content of S and $I(t)$ to measure I Specifically, we let

$$NS(t) = \text{number of susceptibles at time } t,$$
$$NI(t) = \text{number of infectives at time } t$$

The number N is included as a population scale; e.g., it may be the size of the total population at $t = 0$. The functions which describe the contents of the various classes are assumed to be smooth functions of t.

Equations for these functions can be obtained directly from the diagram. In fact, the rate of change of these functions is the "net rate in, less the net rate out". Thus, we obtain the equations

$$dS/dt = A - rIS,$$
$$dI/dt = r[IS(t) - IS(t - \sigma)],$$

for the functions $S(t)$ and $I(t)$ in terms of the parameters A, r and σ.

The basic mechanism for driving a contagious phenomenon is the interaction between susceptibles and infectives. Therefore, the way this interaction is described is very important. Most of the examples studied here model this interaction in the same way that certain chemical reactions are modelled by the *law of mass action*: The rate at which effective contacts occur is taken to be proportional to the number of susceptibles and the number of infectives. Among other things, it is implicit in this assumption that the population is *homogeneously mixing*. That is, every pair of individuals in the population has equal probability of meeting. The mass action law is the continuous analogue of the deterministic Reed–Frost model (see Bailey (1957)).

Several examples will now be listed. They are quite popular, judging from the number of places they have reappeared. It is appropriate to begin this list of examples with the Kermack-McKendrick model, which represents one of the first significant developments in the deterministic theory.

(a) *The Kermack-McKendrick model.* In their papers (1927, 1932, 1933), W. O. Kermack and A. G. McKendrick developed and analyzed various models of the spread of infection. Their basic model is given by

$$S \overset{rIS}{\to} I \overset{qI}{\to} Q.$$

Thus, susceptibles are removed due to the presence of infectives and exposed susceptibles become immediately infective. Infectives are lost through quarantine at a rate proportional to their numbers.

In particular, the population is assumed to be closed in that there is no mechanism for gaining or losing individuals. Therefore, we take N to be the size of the population, and we define

$$NS(t) = \text{number of susceptibles at time } t,$$

$$NI(t) = \text{number of infectives at time } t,$$

$$NQ(t) = \text{number in quarantine at time } t.$$

The fact that the population is closed can be expressed by the equation

$$S(t) + I(t) + Q(t) = 1$$

for all t. Because of this equation, it is necessary only to obtain equations for S and I; however, we may write the full system of equations directly from the diagram

$$\text{(III.1)} \qquad \begin{aligned} dS/dt &= -rIS, \\ dI/dt &= rIS - qI, \\ dQ/dt &= qI \end{aligned}$$

The number r is the contact rate per infective and the number q is the quarantine rate of infectives. These numbers are assumed to be known.

The equations (III.1) make up the Kermack–McKendrick model. In § 2 some of the qualitative properties of solutions are determined. We note here, however, that if $S(0)$, $I(0)$, $Q(0)$ are known, the equations (III.1) have a unique solution for $t \geqq 0$.

(b) *The measles model.* A model for measles was proposed by Soper (1929), and it was developed by Wilson and his co-workers (1942). It has been studied by several investigators since then.

The new features here are that there is a fixed period between exposure and becoming infectious, called the *incubation period*, and there is a fixed period of infectiousness. Thus, rather than an exposed susceptible becoming immediately infectious, it enters the class E, remaining there a fixed period of time. This can be interpreted as a period where the infection incubates in the exposed individual until a sufficient level of infection is acquired. Then, after a fixed period of infectiousness, the individual is removed, in this case through the onset of permanent immunity. The model is described by

$$\overset{A}{\rightarrow} S \overset{rIS}{\rightarrow} E^{\tau} \rightarrow I^{\sigma} \rightarrow R,$$

where A gives the accession of susceptibles (e.g., through migration into the population), τ is the length of the incubation period and gives the holding time in E, and σ is the length of the infectious period. Now the model is not closed, and the full system must be written down. When fixed holding times are involved the differential equations which result directly from the diagram are cumbersome. This is due primarily to questions about how the initial data are to be incorporated in the model. Therefore, it will be convenient to use the following approach to modelling:

For example,

$$E(t) = \begin{array}{l}\text{number of original}\\ \text{members in } E \text{ remain-}\\ \text{ing at time } t\end{array} + \begin{array}{l}\text{number of susceptibles}\\ \text{exposed in the interval}\\ (t-\tau, t).\end{array}$$

We assume that the first term is zero. The number of new exposures occurring in short time dx at time x is given by $rS(x)I(x)\,dx$. Thus, the number of new exposures arising in the previous τ time units is

$$\int_{t-\tau}^{t} rS(x)I(x)H(x)\,dx,$$

where the Heaviside function $H(x)$ is introduced to ensure that only exposures since time $t = 0$ are accounted for (recall $H(x) = 1$ if $x > 0$, $= 0$ if $x < 0$). In this way we see that

$$E(t) = \int_{t-\tau}^{t} rI(x)S(x)H(x)\,dx.$$

Note that those exposed before time $t - \tau$ have become infectious by time t.

A similar analysis will be carried out for I. In fact,

$$I(t) = \begin{array}{l}\text{number of initial}\\ \text{infectives remaining}\\ \text{at time } t\end{array} + \begin{array}{l}\text{number of new}\\ \text{infectives acquired}\\ \text{in time } (t-\sigma, t).\end{array}$$

The first term is assumed to be known and is described by a function $I_0(t)$. Note that $I_0(t) = 0$ for $t > \sigma$ since by time $t = \sigma$, all of the initial infectives will have been removed. The second term can be replaced by

$$\text{number of new exposures occurring in time } (t-\tau-\sigma, t-\tau)$$

since these are the indivduals who become infectious in $(t - \sigma, t)$. Thus,

$$I(t) = I_0(t) + \int_{t-\tau-\sigma}^{t-\tau} rI(x)S(x)H(x)\,dx.$$

Again, as for E, only new infections occurring after time $t - \sigma$ need be accounted for since the others will have been removed by time t.

Finally, we have

$$R(t) = \begin{array}{l}\text{those removed}\\ \text{at time } t = 0\end{array} + \begin{array}{l}\text{new removals}\\ \text{by time } t\end{array} + \begin{array}{l}\text{removed initial}\\ \text{infectives}.\end{array}$$

Obviously, this is

$$R(t) = R(0) + \int_{0}^{t-\tau-\sigma} rI(x)S(x)H(x)\,dx + I_0(0) - I_0(t).$$

Combining these facts, we obtain the full model

$$dS/dt = A - rIS,$$

$$E(t) = \int_{t-\tau}^{t} rI(x)S(x)H(x)\,dx,$$

$$I(t) = I_0(t) + \int_{t-\tau-\sigma}^{t-\tau} rI(x)\,S(x)H(x)\,dx,$$

$$R(t) = R(0) + \int_{0}^{t-\tau-\sigma} rI(x)S(x)H(x)\,dx + I_0(0) - I_0(t).$$

Note that if I and S were known, then E and R could be determined immediately from these formulas. Thus, we need only consider the two equations

$$\text{(III.2)}\qquad \begin{aligned} dS/dt &= A - rIS,\\ I(t) &= I_0(t) + \int_{t-\tau-\sigma}^{t-\tau} rI(x)S(x)H(x)\,dx, \end{aligned}$$

where the contact rate r, the holding times τ and σ, and the initial data $S(0)$ and $I_0(t)$ are assumed known. The equations (III.2) make up the measles model. Certain qualitative features of this model will be found in § 2.

(c) *A relapse-recovery model.* Many phenomena have the property that after a period of infection individuals eventually become susceptible again. A model studied by Hoppensteadt and Waltman (1970, 1971) is described here: In particular, it is given by the compartment diagram

$$S \overset{rIS}{\to} E^{\tau} \to I^{\sigma} \to R^{\omega} \to S,$$

and the equations are derived in a straightforward way as in the previous section. As before, the model reduces to one for determining only S and I The equations for S and I are

$$\text{(III.3)}\qquad \begin{aligned} S(t) &= S(0) - \int_{t-\tau-\sigma-\omega}^{t} rI(x)S(x)H(x)\,dx + [I_0(0) - I_0(t-\omega)]H(t-\omega),\\ I(t) &= I_0(t) + \int_{t-\tau-\sigma}^{t-\tau} rI(x)S(x)H(x)\,dx, \end{aligned}$$

where the contact rate r, the holding times τ, σ, ω and the initial data $S(0)$ and $I_0(t)$ are assumed to be known. The first equation can be derived in the following way:

$$S(t) = \begin{matrix}\text{initial}\\ \text{susceptibles}\end{matrix} - \begin{matrix}\text{susceptibles}\\ \text{exposed by}\\ \text{time } t\end{matrix} + \begin{matrix}\text{susceptibles}\\ \text{exposed by time}\\ t-\tau-\sigma-\omega\end{matrix} + \begin{matrix}\text{initial}\\ \text{infectives}.\end{matrix}$$

The second last term accounts for individuals who have passed through the entire process and have been returned to S. It is given by

$$\int_{0}^{t-\tau-\sigma-\omega} rI(x)S(x)H(x)\,dx$$

while the second term is given by

$$-\int_0^t rI(x)S(x)\,dx$$

Once S and I have been determined, E and R can be found by the formulas

$$E(t) = \int_{t-\tau}^{t} rI(x)S(x)H(x)\,dx,$$

$$R(t) = \int_{t-\tau-\sigma-\omega}^{t-\tau-\sigma} rI(x)S(x)H(x)\,dx + I_0(0) - I_0(t) - [I_0(0) - I_0(t-\omega)]H(t-\omega),$$

where the initial data E_0 and R_0 are assumed to be zero. Finally, we note that the population is closed, and this leads to the equation

$$S(t) + E(t) + I(t) + R(t) = 1$$

for all $t \geqq 0$.

(d) *A threshold of dosage model.* An interesting feature of the models proposed by Cooke (1964) and Hoppensteadt and Waltman (1970, 1971) is the introduction of τ not as a constant, but as a functional of the solution. For example, the dosage that exposed individuals receive over a time interval $(t, t+h)$ may be described by the integral

$$\int_t^{t+h} \rho(x)I(x)\,dx,$$

where ρ is a known function characteristic of the disease, measuring the dosage imparted by each infective in a unit time. Then, the length of time between exposure and infection could be determined from a condition which specifies the dosage: An individual exposed at time $t-\tau$ becomes infectious at time t provided

$$\int_{t-\tau}^{t} \rho(x)I(x)\,dx = m,$$

where m is a given value, called the *threshold of dosage.* This equation then determines the *latency period* as a function of I. This is discussed in greater detail by Waltman (1974).

1.2. A general age dependent theory. The previous section's examples described the dynamics of contagious phenomena acting in a population by keeping track of the total number of susceptibles, infectives, etc It is possible, with a little additional effort, to formulate similar models which contain a great deal more information. These are models which describe the age structure of the various classes, and they include the examples just given.

In this section, a quite general mathematical model will be described and analyzed. The results will be useful in the study of particular models later. Therefore, for the remainder of this section, we shall be concerned with the formulation of a general model

The state of the population with respect to a contagious phenomenon acting in it will be described by a vector $u(c, t)$ each of whose components represents the age distribution of some collection of individuals in the population defined by the phenomenon. For example, in the case of an epidemic disease, which partitions the population into susceptible, infective, quarantined and removed, the state is described by the column vector

$$u(c, t) = \text{col}\,[S(c, t), I(c, t), Q(c, t), R(c, t)],$$

where $S(c, t)$ is the age distribution of susceptibles at time t, etc. In this case, u is a four-dimensional vector, and knowledge of it would give a description of the disease's dynamics. *Age* may have various interpretations here, such as class age or chronological age. In this section, c is taken as a measure of the time since entering the current state, and it is therefore referred to as *class age*. For example, $S(c, t)$ measures the number of individuals at time t whose age in S is c. Chronological age may also be accounted for by a straightforward modification of this model which adds a new variable a.

In general, the state vector will consist of n components, each giving the age distribution of some class in the population. The model to be presented here is a system of equations for the state vector. It is based on the ideas used in the equation of age dependent growth, but it is more general. In particular, the removal rates are now allowed to depend on the state.

We consider an initial-boundary value problem for the n-vector $u(c, t)$:

$$\begin{aligned} \partial u/\partial t + \partial u/\partial c &= F(c, t, u(\cdot, t))u(c, t), && c, t \geqq 0, \\ u(0, t) &= G(t; u(\cdot, t)), && t > 0, \\ u(c, 0) &= u_0(c), && c \geqq 0, \end{aligned} \tag{III.4}$$

where the functionals F and P have special forms. The $n \times n$-matrix F is given by

$$F_{i,j}(c, t, u(\cdot, t)) = C_{i,j}(c, t) - \eta_{i,j}(c, t, u)$$

for $i, j = 1, \cdots, n$, where $C_{i,j}$ are real numbers and $\eta_{i,j}$ are given by

$$\eta_{i,j}(c, t; u) = \int_0^\infty \eta_{i,j}(c, t, c')u(c', t)\,dc'$$

We will denote by η, the $n \times n$-matrix

$$\eta(c, t; u) = \{\eta_{i,j}(c, t; u)\}_{i,j=1}^{n}. \tag{III.5}$$

The n-vector G has the form

$$G(t; u(\cdot, t)) = g(t) + \int_0^\infty [\beta(c, t) + \gamma(c, t)C(c, t) + \varepsilon(c, t)\eta(c, t; u)]u(c, t)\,dc,$$

where g is an n-vector and β, γ and ε are $n \times n$-matrices. While the functionals F and G seem rather special, they are actually typical of the models encountered,

The matrix C accounts for relative removal and growth rates in the various classes, and the collection of row vectors $\{\eta_{ij}\}_{j=1,\cdots,n}$ accounts for removals from the ith class due to the presence of the other classes. On the other hand, the terms making up G have the following meanings: The vector g accounts for growth of population through migration, and $\beta(c, t)$ plays the role of a fertility matrix. It accounts for births within the various classes. Finally, the matrices γ and ε distribute the removals among the other classes where they enter with age zero.

Since it is important for the state vector to have only nonnegative components, the notation $u \geqq 0$ will be used in instances where each component u_i of u is nonnegative. Thus, $u \geqq 0$ if and only if $u_i \geqq 0$ for all $i = 1, \cdots, n$. When $u \geqq 0$, the vector u is referred to as being nonnegative.

The main result we shall use about (III.4) is given now.

THEOREM. *Suppose the following hypotheses* (H1–H5) *are satisfied*:

(H1) *$u_0 \in R^n$ is a nonnegative function for $0 \leqq c < \infty$, having bounded support (i.e., there is a constant $c^* > 0$ such that $u_0(c) = 0$ for $c > c^*$), and $g \in R^1$, β, γ, ε, $C \in R^{n \times n}$ and $\eta_{ij} \in R^{1 \times n}$ are continuous functions of their arguments for $0 \leqq c$, c', $t < \infty$.*

(H2) *For any integrable function $v(c) \in R^n$ having bounded support and satisfying $v(c) \geqq 0$, the matrix $\eta(v)$ defined in* (III.5) *is positive semi-definite (i.e., the quadratic form $\xi^{tr}\eta(v)\xi \geqq 0$ for all $\xi \in R^n$).*

(H3) *For any v as in* (H2), *the off-diagonal terms in the matrix $F(c, t; v(\cdot))$ are nonnegative for $0 \leqq c, t < \infty$*

(H4) *For any v as in* (H2), *$G(t; v(\cdot)) \geqq 0$ for all $t > 0$.*

(H5) For $j = 1, \ldots, n$,

$$\sum_{i=1}^{n} (\varepsilon_{ij} - \delta_{ij}) \leqq 0,$$

where $\delta_{ij} = 0$ if $i \neq j$, and $= 1$ if $i = j$.

If these conditions are satisfied, the problem (III.4) *has a unique solution for $0 \leqq c, t < \infty$. Moreover, this function has the following properties:*

(C1) *$u(c, t) \geqq 0$ for all $0 \leqq c, t < \infty$.*

(C2) *$u(c, t)$ is continuous for $c \neq t$ and continuously differentiable along the characteristic lines $c = t +$ const*

(C3) *$u(c, t) = 0$ for $c \geqq c^* + t$.*

Briefly, the hypotheses of this theorem have the following significance: (H1) specifies the smoothness required of the data and (H2) guarantees that the growth of the solution along the characteristic lines is dominated by the linear part of the equation (i.e., by $C(c, t)u(c, t)$); this plays an important role in obtaining a priori bounds for the solution. Conditions (H3) and (H4) ensure that the solution is nonnegative. And (H5) plays an important role in obtaining a bound on the birth rate, $u(0, t)$. The proof of the theorem and special cases of the general model are given in Hoppensteadt (1974b).

1.3. Nonhomogeneous mixing: a model of spatial spread. The problems discussed in the earlier sections have focused on epidemics which take place at a point.

They are based on an assumption that all candidates in various classes have equal probability of meeting; i.e., there are no individuals in a given class which are distinguished from their colleagues. When this occurs, the population is said to be homogeneously mixing. One way of getting around this restriction is to break the population down into subpopulations in which there is homogeneous mixing, and then consider the interactions between these subpopulations. The total population can then be broken into a finite number of homogeneously mixing groups. It is useful to extend this idea by introducing a continuous decomposition of the population. This idea is illustrated through a model for spatial spread of an infection: The population is considered to be dispersed over a planar region $\Omega \subset R^2$, and it is labelled according to its position with respect to some fixed set of coordinates.

Consider some contagious phenomenon acting which partitions the population at each point into susceptible, infective, and quarantined individuals. These classes will then be described by age-space distributions:

$$S(c, t, P) = \text{age-space distribution of susceptibles at time } t,$$

$$I(c, t, P) = \text{age-space distribution of infectives at time } t,$$

$$Q(c, t, P) = \text{age-space distribution of quarantinees at time } t.$$

Thus, the number of susceptibles in the age bracket $(c, c + dc)$ residing in the neighborhood A is

$$N \int_c^{c+dc} \int_A S(c', t, P)\, dP\, dc'.$$

The epidemic described is of type

$$S \to I \to Q,$$

and the transitions are given by the following assumptions

(i) The relative rate of exposure of susceptibles at P is given by

$$\int_{R^2} \int_0^{\infty} \xi(c, t, P; c', P') I(c', t, P')\, dc'\, dP'.$$

The function $\xi(c, t, P, c', P')$ measures the influence of infectives of age c' at P' on exposing susceptibles of age c at P.

(ii) The relative rate of quarantine of infectives at P is given by $q(c, t, P)$.

From these assumptions, we deduce the model

$$\left(\frac{\partial}{\partial t} + \frac{\partial}{\partial c}\right)\begin{pmatrix} S(c, t, P) \\ I(c, t, P) \\ Q(c, t, P) \end{pmatrix} = \begin{pmatrix} -\int_{R^2}\int_0^{\infty} \xi(c, t, P; c', t') I(c', t, P')\, dc'\, dP', & 0 & 0 \\ 0 & -q & 0 \\ 0 & 0 & 0 \end{pmatrix}\begin{pmatrix} S \\ I \\ Q \end{pmatrix}$$

The associated initial conditions are

$$(S(c, 0, P), I(c, 0, P), Q(c, 0, P)) = (S_0(c, P), I_0(c, P), 0),$$

and the boundary conditions are

$$(S(0,t,P), I(0,t,P), Q(0,t,P))$$

$$= \left(0, \int_0^\infty S(c,t,P) \int_{R^2} \int_0^\infty \xi(c,t,P,c',P') I(c',t,P')\, dc'\, dP'\, dc, \right.$$

$$\left. \int_0^\infty q(c,t,P) I(c,t,P)\, dc \right)$$

for $t > 0$ The existence theorem in § 1.2 can be extended to the present case under the conditions stated here.

Some qualitative properties of the solutions of the model are given in § 2.

It is important to think of this example as a model illustrating a general indexing of subpopulations. Here the index P can be physically interpreted as the location of the subpopulation It is easy to see how a model of a "completely" nonhomogeneously mixing population can be formulated · P in that case can be taken as an indexing of the members of the population. Thus, the table of values of the function $S(c, t, P)$ lists all the individuals who are susceptible at (c, t) by name.

2. Qualitative behavior of deterministic epidemics. Three basic types of behavior are studied in this chapter. Each one involves substantially different methods of analysis. In § 2 1, an analysis is made of the sensitivity of solutions to various parameters in the models. The result described there is a fairly general threshold theorem which includes the threshold theorems of Kermack-McKendrick, Landau-Rapoport, Marchand and L. O Wilson. Moreover, the result is developed for age dependent models.

Section 2.2 contains an analysis of the oscillations and stability for solutions of the relapse-recovery model. In particular, it is shown how the presence of a threshold can act as a mechanism for driving oscillations.

The remaining section of this chapter studies certain properties of models for the spatial spread of infection. Section 2.3 begins with some general considerations; then, a diffusion approximation is obtained and the existence of various types of solutions, such as progressing wave solutions and domains of constant infection are studied

2.1. The threshold theorem. One of the most important contributions of epidemic models has been in showing which are the principal parameters governing the spread of contagion This has been a focal point of many investigations, but the work of Kermack and McKendrick is the outstanding one. It provided one of the first results about threshold values for population size

The threshold theorem described here is substantially more general than Kermack and McKendrick's result, yet, certain technical points aside, it rests on basically the same argument.

The model is an age dependent one of type

$$S \to I \to$$

Susceptibles are removed through exposure to infectives, but the measure of infection depends only on the age of infectives, neither on the age of susceptibles exposed nor on time. Furthermore, there is a nondecreasing function $\tau(t)$ such that susceptibles exposed at $\tau(t)$ become infectious at time t. Thus, $t - \tau(t)$ measures the incubation time. Finally, infectives are removed at an age specific rate (either through death or quarantine) until age σ at which the period of infectiousness expires. Thus, the model can be written as

$$
\begin{aligned}
&\frac{\partial S}{\partial t} + \frac{\partial S}{\partial a} = -\left[\int_0^\sigma r(a')I(a', t)\,da'\right]S, \\
&S(0, t) = 0, \quad S(a, 0) = S_0(a), \\
&\partial I/\partial t + \partial I/\partial a = -\alpha(a)I, \\
&I(0, t) = -(d/dt)\int_0^\infty S(a, \tau(t))\,da, \quad I(a, 0) = I_0(a).
\end{aligned} \tag{III.6}
$$

The equation for $I(0, t)$, the birth rate of infectives, is obtained by observing that the total rate at which susceptibles are exposed at time $\tau(t)$ is

$$(d/dt)\int_0^\infty S(a, \tau(t))\,da,$$

since the size of the susceptible population at $\tau(t)$ is

$$\int_0^\infty S(a, \tau(t))\,da$$

and the only change in this is due to exposure to infection

The critical parameter in this analysis will be the number of susceptibles expected to be infected by each infective. If this number is less than one, then it might be expected that the infection would die out of the population after affecting only a number of susceptibles comparable to the number of initial infectives. On the other hand, if this parameter is greater than one, a snowballing effect is to be expected and a significant epidemic will result.

The number γ of initial susceptibles expected to be exposed to each infective is given by the formula

$$\gamma = S_0 \int_0^\sigma r(a)\exp\left[-\int_0^a \alpha(a')\,da'\right]da,$$

where

$$S_0 = \int_0^\infty S_0(a)\,da$$

is the total number of initial susceptibles and the second factor in γ measures the infective capability of each infective. The function

$$\exp\left[-\int_0^a \alpha(a')\,da'\right]$$

gives the probability of an initial infective surviving to age a and $r(a)$ measures the effectiveness of those surviving to age a.

The value $\gamma = 1$ will be shown to be a critical value in that the final size of the epidemic changes dramatically as γ passes through 1. Because of this, the value one is called a *threshold value* for γ.

In fact we shall show that the limit

$$\lim_{t \to \infty} \int_0^\infty S(a, t)\, da = S_\infty$$

exists, and then describe its dependence on γ.

From integrating the problem for I, we have

$$I(a, t) = \begin{cases} I_0(a - t) & \text{for } a > t, \\ I(0, t - a) \exp\left[-\int_0^a \alpha(a')\, da' \right] & \text{for } a < t, \end{cases}$$

where $I_0(a, t)$ can be determined in terms of α, I_0 and r. Since r vanishes for $a > \sigma$, we see that $I_0(a, t) = 0$ for $a > \sigma$.

Let us consider the total susceptible population at time t

$$S(t) = \int_0^\infty S(a, t)\, da.$$

Since $S(0, t) = 0$,

$$S(t) = \int_t^\infty S(a, t)\, da.$$

If $S_0(a)$ has bounded support, then so does $S(a, t)$ for each $t > 0$. Therefore,

$$\frac{dS}{dt} = \int_0^\infty \frac{\partial S}{\partial t}(a, t)\, da = -\left(\int_0^\sigma r(a) I(a, t)\, da \right) S(t),$$

$$S(0) = \int_0^\infty S(a, 0)\, da = \int_0^\infty S_0(a)\, da \equiv S_0$$

It follows that $\lim_{t \to \infty} S(t) = S_\infty$ exists as a number between 0 and S_0, and

$$S_\infty / S_0 = \exp\left[-\int_0^\infty \int_0^\sigma r(a) I(a, t')\, da\, dt' \right].$$

The next step involves relating the integral in the exponent to the number S_∞ / S_0.

Now,

$$\int_0^\sigma r(a) I(a, t)\, da = \int_t^\sigma r(a) I_0(a, t)\, da + \int_0^t r(a) I(0, t - a) \exp\left(-\int_0^a \alpha \right) da.$$

Recall that $r(a) = 0$ for $a > \sigma$ so the first integral on the right vanishes for $t > \sigma$. It follows that

$$\int_0^t \int_0^\sigma r(a)I(a,t')\,da\,dt'$$
$$= m(t) - \int_0^t \int_0^{t'} r(a)\exp\left(-\int_0^a \alpha\right)\frac{d}{dt'}\int_0^\infty S(a',\tau(t'-a))\,da'\,da\,dt',$$

where

$$m(t) = \int_0^t \int_{t'}^\sigma r(a)I_0(a,t')\,da\,dt'.$$

Interchanging the order of integration in the second integral leads to the formula

$$\int_0^t \int_0^\sigma r(a)I(a,t')\,da\,dt'$$
$$= m(t) - \int_0^t r(a)\,e^{-\int_0^a \alpha}\left\{\int_0^\infty S(a',\tau(t-a))\,da' - S_0\right\}da$$

Hence, for $t > \sigma$,

$$S(t) = S_0 \exp\left[\int_0^\sigma r(a)\,e^{-\int_0^a \alpha}\,S(\tau(t-a))\,da\right].$$

Passing to the limit $t = \infty$ in this formula, we obtain

$$S(\infty) = S_0 \exp\left\{S(\tau(\infty))\int_0^\sigma r(a)\,e^{-\int_0^a \alpha}\,da - m(\infty) - S_0\int_0^\infty r(a)\,e^{-\int_0^a \alpha}\,da\right\},$$

where $0 \leqq \tau(\infty) \leqq \infty$. Recalling that $r(a) = 0$ for $a > \sigma$, we have that

$$\frac{S(\infty)}{S_0} = \exp\left\{S_0\int_0^\sigma r(a)\,e^{-\int_0^a \alpha}\,da\left[\frac{S(\tau(\infty))}{S_0} - 1\right.\right.$$
$$\left.\left.-\left(\int_0^\sigma \int_t^\sigma r(a)I_0(a,t)\,da\,dt \Big/ S_0\int_0^\sigma r(a)\,e^{-\int_0^a \alpha}\right)\right]\right\}$$

Since $S(t)$ is a nondecreasing function of t, there is a number v given by the formula

$$S(\tau(\infty)) = e^v S(\infty).$$

Let

$$F = e^v S(\infty)/S_0;$$

then the equation for F resulting from the above analysis is

$$F = \exp[\gamma(F - 1 - \varepsilon)], \tag{III.7}$$

where

$$\gamma = S_0\int_0^\sigma r(a)\exp\left(-\int_0^a \alpha\right)da$$

and

$$\varepsilon = (1/\gamma)\left[\int_0^\sigma \int_t^\sigma r(a) I_0(a,t)\, da\, dt - \nu\right].$$

Thus, we see that the parameter γ discussed at the beginning of this section arises naturally in the calculation of the final size. The number ε measures the infectiousness of the initial infectives. In particular, if $\tau(\infty) = \infty$, then $\nu = 0$, so ε measures the infective capability of the initial infectives relative to the number of susceptibles expected to be exposed to each infective.

If $0 < \tau(\infty) < \infty$, then it must be that

$$\nu < \int_0^\sigma \int_t^\sigma r(a) I_0(a,t)\, da\, dt.$$

In fact, for ν equal to this value, we have

$$S(\infty) = S_0 e^{-\gamma},$$

so $S(\tau(\infty)) = S_0$, or $\tau(\infty) = 0$. While for ν greater than this quantity, equation (III.7) shows there are two values for $S(\tau(\infty))$ which is impossible. Therefore, we need only consider the solution of (III.7) taking $\varepsilon > 0$.

In order to display most clearly the solutions of (III.7), we consider the roots as functions of the parameter γ. The analysis is simplified if we consider first the case where $\varepsilon = 0$. In this case, there are two real roots for F: One is obviously $F = 1$, the other is described by the solid line in Fig. III.1. For $\varepsilon > 0$ but small,

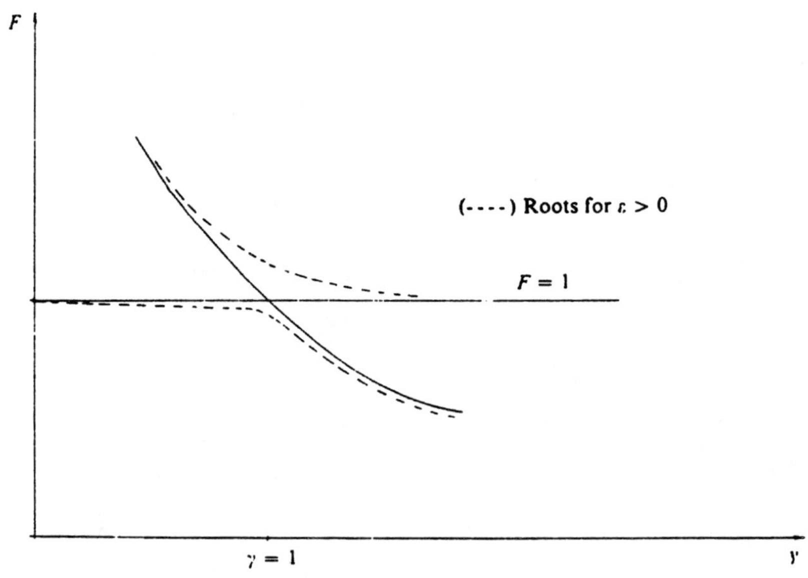

FIG III 1

these roots split (or bifurcate) as described in Fig. III.1 by the dotted lines. Thus, for $\varepsilon > 0$, there is a unique root satisfying

$$F \leqq 1$$

for each γ. Of course, this is the only root of interest.

This analysis was described for age independent models by Hoppensteadt (1974), where equation (III.7) also arises. A numerical evaluation of the roots of (III.7) was carried out there and the results were described by a graph which is reproduced here in Fig. III.2. As expected, the final size of the susceptible population decreases, and the threshold value $\gamma = 1$ loses significance for large ε. In particular, the value $\varepsilon = 0.1$ corresponds roughly to there being one infective to every ten susceptibles, which is rather extreme.

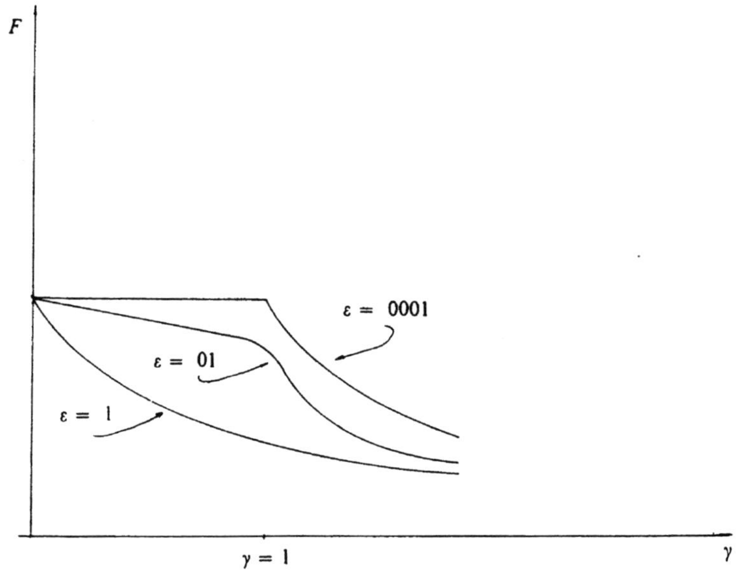

FIG III 2

This result gives only a qualitative description of the final size. If the number ν was known, then quantitative results could be obtained. In the model described in § 1.1, the function $\tau(t)$ is given by $\tau(t) = t - \tau$, where τ is a fixed constant. Hence, for that model, $\tau(\infty) = \infty$ and so the threshold result can be applied there with $\nu = 0$.

The results of this section are summarized in the following theorem.

THRESHOLD THEOREM. *Let $\tau(t)$ be a nondecreasing function of t with $\tau(0) = 0$. If $S(a, t)$ and $I(a, t)$ are determined by the system* (III.6) *where α and r are continuous nonnegative functions, then the final size of the epidemic is given by the limit*

$$\lim_{t \to \infty} \int_t^\infty S(a, t)\, da = S_\infty .$$

Moreover, for initial data such that

$$\varepsilon \ll 1,$$

the value S_∞ considered as a function of the parameter

$$\gamma = S_0 \int_0^\sigma r(a) \exp\left[-\int_0^a \alpha\right] da$$

decreases gradually for $0 \leqq \gamma \leqq 1$, but it changes significantly for $\gamma > 1$. Specifically, the value of $e^\gamma S(\infty)/S_0$ is given by the admissible root of (III.7).

This result was derived by Kermack and McKendrick for their model (1927), by Landau and Rapoport (1951) and Marchand (1955) for the measles model, and by Heathcote (1971) and Wilson (1972) for the threshold of dosage model. The first two of these results are of interest to us here since they will be used later.

For the Kermack-McKendrick result, we take r and α to be independent of a and $\sigma = \infty$, $\tau(t) = t$. Thus,

$$\gamma = S_0 r/\alpha.$$

For the measles model, we take r independent of a, $\alpha = 0$ and $\tau(t) = t - \tau$. Thus,

$$\gamma = S_0 r \sigma.$$

We shall see in later sections that these thresholds can act as mechanisms for driving oscillations in the models. These observations will be based on a slightly different interpretation of the threshold. Namely, since the threshold value for γ is one, we may say, alternatively, that the value

$$S^* = 1 \Big/ \int_0^\sigma r(a) \exp\left(-\int_0^a \alpha(a')\, da'\right) da$$

is the threshold value for the initial population size. If $S_0 > S^*$, then $\gamma > 1$, while if $S_0 < S^*$, $\gamma < 1$. This was the interpretation given by Kermack and McKendrick.

Threshold results for more complicated diseases have been studied. For example, MacDonald, Nasell and Hirsch and Stirzaker have investigated the role played by thresholds in schistosomiasis.

Finally, a note of caution. Real data must be consulted before conclusions about qualitative features, such as thresholds, can be drawn. This is illustrated by the example of how thresholds have been used incorrectly to describe oscillations in the recurrence of measles epidemics. Measles has an alternate low year, high year oscillation, and so has a period of two years. It was suggested that a threshold was responsible; in particular, that every other year the susceptible population would greatly exceed its threshold value since after a low year, more susceptibles would be available. However, chickenpox is a disease quite similar to measles in terms of contagion, and chickenpox epidemics of comparable size occur every year. Analysis of simple mathematical models suggested that the threshold could not account for the difference in periodicity, and data analysis suggested that a nonconstant contact rate r should be used. The measles model with a periodic contact rate was analyzed by Stirzaker (1974b). He showed that the parameter values for chickenpox were correct to create ordinary resonance in the model, and so solutions of period one year. Also, the parameter values for measles were correct to create parametric resonance, and so solutions having period twice that of the contact rate.

2.2. Relapse-recovery model. Many contagious phenomena have the feature that susceptibles who become infectious may eventually become susceptible again. For example, such phenomena can arise in infectious diseases such as gonorrhea, and in studies of criminology, e.g., where susceptibles may represent potential criminals, infectives active criminals, and removed incarcerated criminals.

A model for relapse-recovery phenomena was proposed and studied by Hoppensteadt and Waltman (1971). A simple version of that model will be analyzed in this section. The model postulated here is of type $S \to I \to S$ and is for a phenomenon acting in a closed population. It is based on the following assumptions:

(i) The relative exposure rate of susceptibles is given by rI, r a known constant.

(ii) Once exposed, individuals become immediately infectious and after a fixed infectious period of length σ, they are removed.

(iii) Removed individuals remain so only for a fixed period of length ω

As before, these assumptions can be shown to give the following equations for $S(t)$ and $I(t)$ in terms of the initial number of susceptibles (S_0), the dynamics of the initial infectives ($I_0(t)$) and the data r, σ and ω:

$$S(t) = S_0 - \int_{t-\sigma-\omega}^{t} (rIS)(x)\,dx + I_0(t-\omega)H(t-\omega),$$

$$I(t) = I_0(t) + \int_{t-\sigma}^{t} (rIS)(x)H(x)\,dx.$$

It follows from the theorem for the general model that the solution of this problem exists for all $0 \leqq t < \infty$ and that $I(t)$ satisfies

$$0 \leqq I(t) \leqq I(0) + S_0$$

for all $0 \leqq t < \infty$.

The main feature of this epidemic can be seen by considering a slight simplification where we take $\omega = 0$. Also, a new time variable is convenient; let $t \to \sigma t$. In this case, we may use the fact that the population is closed ($S(t) + I(t) = 1$) to reduce the model to the one equation

$$I(t) = I_0(t) + r\sigma \int_{t-1}^{t} I(x)[1 - I(x)]\,dx. \tag{III.8}$$

We shall study the long time behavior of $I(t)$, and particularly interesting questions arise about the existence of periodic solutions and the approach of solutions to steady states.

One might suspect that periodic solutions do exist. In fact, we saw in the previous section that in the case $\omega = \infty$, there is a threshold value for the susceptible population, $S_0 = 1/r\sigma$. If the susceptible population begins above this threshold, an epidemic will occur in which the susceptible population is driven below its threshold value. Once near S_∞, few additional susceptibles are exposed. With the mechanism described for reintroduction of susceptibles, we see that the susceptible population may automatically rise above the threshold again. When this happens, another epidemic will take place driving the susceptible population back below threshold. Therefore, there is a mechanism in the model for driving the oscillations. However, we can see from this argument that the epidemics are probably not periodic functions: Since the renewal of susceptibles does not occur instantaneously, as soon as the susceptible population rises above $(1/r\sigma)$, it is acted upon by infectives. Therefore, it is expected that the susceptible population will not rise again to S_0. Since this will happen in each cycle, the maxima of $S(t)$ should form a decreasing sequence. Similarly, the minima of S should form an increasing sequence.

This discussion suggests that the solution of (III.8) approaches a constant value. To determine candidates for this value, let us suppose $I(t)$ has a limit I_∞, as $t \to \infty$. I_∞ must satisfy

$$I_\infty = r\sigma I_\infty(1 - I_\infty).$$

Thus, the possible limits are $I_\infty = 0$ and $I_\infty = (r\sigma - 1)/r\sigma$. The first limit shows that $S(\infty) = 1$, which says that the infection dies out of the population. The second limit shows that $S(\infty) = 1/r\sigma$, the threshold value for the susceptible population. Thus, the threshold value gives an endemic size for the infection. We shall now study the stability of these two states.

This problem has been investigated by Cooke and Yorke (1973) and by Greenberg and Hoppensteadt (1974), and it has been shown that the solution of the problem always approaches a steady state as $t \to \infty$. The proof of this result will only be partially described; the case $r\sigma \leqq 3$ illustrates the three kinds of behavior exhibited by the solution.

First, we consider a difference equation associated with (III.8):

$$I_{n+1} = r\sigma I_n(1 - I_n).$$

This recursion scheme represents iteration of the quadratic $r\sigma I(1 - I)$, and there are three typical cases, as illustrated in Fig. III.3(a), (b), (c).

The solutions of the difference equation can be related to the solution of (III.8) in the following way: For $r\sigma < 1$, given $I_0(t)$, there is a sequence of iterates of the difference equation $\{I_n\}$ such that

$$\sup_{[n,n+1]} I(t) \leqq I_n.$$

Thus, all solutions of (III.8) approach zero as $t \to \infty$ if $r\sigma < 1$. For $2 > r\sigma > 1$, given $I_0(t) \neq 0$, there is a sequence of iterates $\{I_n\}$ converging to $1 - (1/r\sigma)$ such that

$$\sup_{[n,n+1]} |I(t) - 1 + (1/r\sigma)| \leqq |I_n - 1 + (1/r\sigma)|.$$

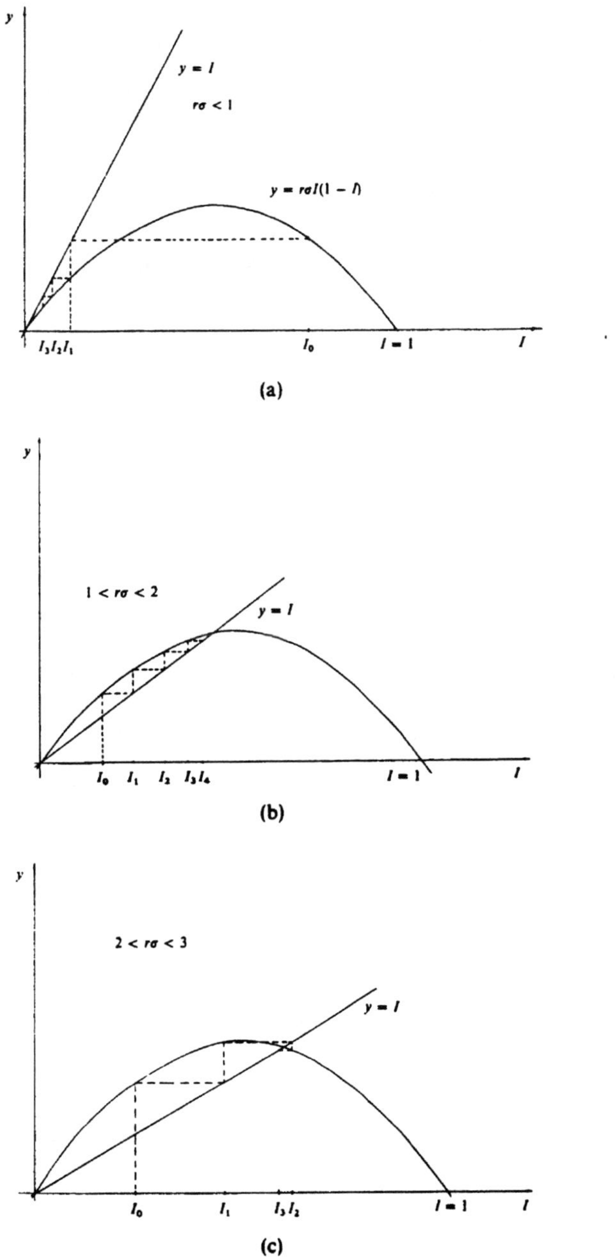

FIG III 3 (a) *illustrates how all solutions of the difference equation converge monotonically to zero for* $r\sigma < 1$ (b) *shows that for* $1 < r\sigma < 2$, *all solutions with* $I_0 \neq 0$ *eventually converge monotonically to the steady state* $1 - (1/r\sigma)$. *Finally,* (c) *demonstrates that for* $2 < r\sigma < 3$, *the solution eventually approaches the nontrivial steady state in an oscillatory way For* $r\sigma \gg 3$, *the rest state is repulsive. The difference in scale between the three figures is indicated by the line* $y = I$

Finally, for $2 < r\sigma < 3$, there are two sequences I_n^u and I_n^l which converge to $1 - (1/r\sigma)$ with

$$I_n^u = r\sigma I_n^l(1 - I_n^l), \quad I_{n+1}^l = r\sigma I_n^u(1 - I_n^u),$$

such that

$$I_n^l - 1 + (1/r\sigma) \leqq I(t) - 1 + (1/r\sigma) \leqq I_n^u - 1 + (1/r\sigma),$$

for $n \leqq t \leqq n + 1$. This method of proof breaks down for $r\sigma \gg 3$, but it can be shown by other methods that all nontrivial solutions approach $1 - (1/r\sigma)$ in an oscillatory way for $r\sigma \geqq 3$.

2.3. Models of spatial spread. Methods for analyzing spatial spread of infection will be described in this section for the spatial analogue of the Kermack-McKendrick model. Because of our interest in studying the long time behavior of this problem, we may as well restrict ourselves to the age independent version; namely,

$$\partial S/\partial t = -\left[\int_{R^2} \xi(P, P')I(t, P')\,dP'\right]S(t, P),$$

$$\partial I/\partial t = \left[\int_{R^2} \xi(P, P')I(t, P')\,dP'\right]S(t, P) - q(t, P)I(t, P),$$

$$\partial Q/\partial t = q(t, P)I(t, P).$$

The functions S, I and Q give the spatial distribution of the susceptibles, infectives and quarantined populations; $\xi(P, P')$ measures the influence of infectives at P' in exposing susceptibles at P, and $q(t, P)$ measures the relative quarantine rate of infectives at P.

(a) *The final size distribution.* The total size of the epidemic can be computed as for the Kermack-McKendrick model In fact, let us take $q(t, P) = q(P)$, to be independent of time. Then

$$S(t, P) = S(0, P)\exp\left[\int_0^t \int_{R^2} p(P, P')(\partial Q/\partial t)(P', t)\,dP'\,dt\right],$$

where

$$p(P, P') = \xi(P, P')/q(P').$$

From the third equation of the model, we see that Q is a nondecreasing function of t. Moreover, we have from the model that

$$\frac{\partial}{\partial t}(S(t, P) + I(t, P) + Q(t, P)) = 0,$$

hence,

$$S(t, P) + I(t, P) + Q(t, P) = N(P),$$

the initial population distribution Since each of these functions is nonnegative,

$Q(t, P)$ is bounded independently of t. This shows that

$$\lim_{t\to\infty} Q(t, P) = Q^*(P)$$

exists for all $P \in R^2$ and that $(\partial Q/\partial t)(t, P)$ approaches zero. On the other hand, the model shows that

$$\frac{\partial Q}{\partial t} = q(P)\left\{N(P) - Q(t, P) - S(0, P)\exp\left[\left[\int_{R^2} p(P, P')Q(t, P')\,dP'\right]\right]\right\}$$

and passage to the limit $t = \infty$ gives

$$N(P) = Q^*(P) + S(0, P)\exp\left[\left[\int_{R^2} p(P, P')Q^*(P')\,dP'\right]\right].$$

This equation can be solved for the final size distribution of removals provided p satisfies certain smoothness conditions (see Yosida (1960)). Finally, since $I(\infty, P) = 0$, we can recover $S(\infty, P)$ from the formula $S(\infty, P) = N(P) - Q^*(P)$.

(b) *A diffusion approximation.* A question of great interest for models of spatial spread has to do with the existence of wave-like phenomena. Certain epidemics have been observed proceeding from a central point across the countryside in a wave-like way (Bailey (1957)). This suggests that such waves should be predictable from the model.

Now, let us assume for simplicity that

$$\xi(P, P') = k(P - P') \quad \text{and} \quad q = \text{const.}$$

Thus, the influence of infectives on susceptibles is a function only of distance between them and not on specific preferred directions. The model then becomes

$$\begin{aligned}\partial S/\partial t &= -S(t, P)\int_{R^2} k(P - P')I(t, P')\,dP',\\ \partial I/\partial t &= S(t, P)\int_{R^2} k(P - P')I(t, P')\,dP' - qI(t, P),\\ \partial Q/\partial t &= qI(t, P).\end{aligned} \tag{III.9}$$

By making two additional assumptions about the problem, this model can be approximated by a nonlinear diffusion equation. Let us suppose that $k(-P) = k(P)$, that $k(P) = 0$ for $|P| \geqq \delta$, where δ is some small number, and that the function $I(t, P)$ does not change too drastically over any set of radius δ: For example, let us suppose the third derivatives of I with respect to the components of P are $O(1)$ on any such set. Then the integral

$$\int_{R^2} k(P - P')I(t, P')\,dP' = \int_{R^2} k(P')I(t, P - P')\,dP'$$

may be rewritten using Taylor's formula,

$$\begin{aligned}I(t, P - P') = I(t, P) - (P' \cdot \nabla_P)I(t, P) + (1/2)(P' \cdot \nabla_P)^2 I(t, P)\\ -(1/6)(P' \cdot \nabla_{P''})^3 I(t, P'')|_{P'' = P + P^*},\end{aligned}$$

where P^* is some point between P and P'. With this, we obtain

$$\int_{R^2} k(P')I(t, P - P')\, dP' = I(t, P) \int_{R^2} k - \int_{R^2} k(P')(P' \cdot \nabla_P)\, dP' I(t, P)$$
$$+ \left(\frac{1}{2}\right)\left[\int_{R^2} k(P')(P' \cdot \nabla_P)^2\, dP'\right] I(t, P)$$
$$- \left(\frac{1}{6}\right)\int_{R^2} k(P' \cdot \nabla_{P''})^3\, dP' I(t, P'')\bigg|_{P'' = P + P^*}$$

Since $k(-P) = k(P)$, the second integral on the right side of this equation vanishes. Also, since the third derivatives of I are not large, the fourth integral on the right side is of order δ^3 and will therefore be neglected. This argument explains why we may replace the integral by the differential operator

$$\alpha I + AI,$$

where

$$AI = \sum_{i,j=1}^{2} \beta_{ij}\left(\frac{\partial^2 I}{\partial p_i\, \partial p_j}\right), \quad P = (p_1, p_2), \quad \alpha = \int_{R^2} k(P)\, dP,$$
$$\beta_{ij} = \int_{R^2} k(P')p_i' p_j'\, dP', \qquad i, j = 1, 2.$$

The diffusion approximation to the model is

$$\partial S/\partial t = -S[\alpha I + AI],$$
$$\partial I/\partial t = S[\alpha I + AI] - qI,$$
$$\partial Q/\partial t = qI.$$

(c) *Domain of constant infection.* One approach to the detection of wave-like behavior for this model is to seek *domains of constant infection.* In this section, we consider the two-dimensional diffusion approximation where we assume for simplicity that

$$\int_{R^2} k(P')p_i p_j\, dP' = 0 \quad \text{if} \quad i \neq j \quad \text{and} \quad \int_{R^2} k(P')p_i'^2\, dP' = \beta \quad \text{for} \quad i = 1, 2.$$

Thus, the operator A is simply the Laplacian. This assumption simplifies the analysis substantially although a similar one can be carried out for the general case.

The approach taken here is similar to the one described by Fisher (1936) and Bailey (1957), and it has already been used in Part II, § 4.2, to study the spread of a gene in a spatially distributed population. The method involves finding a function $R(t)$ such that

$$\int_{|P| \geq R(t)} I(t, P)\, dP = \text{const.}$$

This integral gives the size of the infective population residing outside the circle $|P| = R$. Thus, the problem is to determine how this circle grows with time. If the constant is small, we may assume that the susceptible population remains constant outside this circle. In addition, let us suppose that the infectives are initially concentrated at $P = 0$:

$$I(0, P) = I_0\, \delta(P).$$

With these assumptions, the problem reduces to the linear parabolic equation

$$\partial I/\partial t = (\alpha S - q)I + \beta S \nabla_p^2 I$$

with the initial condition

$$I(0, P) = I_0\, \delta(P).$$

To ensure that the population can support an epidemic at each point, we suppose that the susceptible population initially exceeds its threshold value for each P. Thus, $S > q/\alpha$. Setting

$$J(t, P) = I(\tau/\beta S, P)\, e^{-\mu\tau},$$

where $\tau = \beta S t$, $\mu = (\alpha S - q)/\beta S$, we see that J satisfies

$$\partial J/\partial t = \nabla_P^2 J, \quad J(0, P) = I_0\, \delta(P).$$

The solution of this problem is (see Courtant-Hilbert (1959), p. 198)

$$J(t, P) = (I_0/4\pi\tau) \exp\,[-|P|^2/4\tau]$$

or

$$I(t, P) = (I_0/4\pi\beta S t) \exp\,[(\alpha S - q)t - (|P|^2/4\beta S t)].$$

If the size of the infective population outside the circle is taken to be C, then

$$\begin{aligned} C &= (I_0/\pi)\, e^{(\alpha S - q)t} \int_{|P| \geqq R} e^{-|P|^2/4\beta S t}/4\beta S t \\ &= I_0 \exp\,[(\alpha S - q)t - (R^2/4\beta S t)]. \end{aligned}$$

Solving this equation for R, we see that the radius for the circle is

$$R(t) = 2t\sqrt{\beta S(\alpha S - q)} + o(t) \quad \text{as} \quad t \to \infty.$$

The result is that the circle of constant infection grows at a rate which is asymptotically independent of C.

(d) *Progressing wave solutions for a simple epidemic in one dimension.* Another investigation of wave-like phenomena was made by D. Mollison (1972). We shall illustrate his result by considering (III.9) with $q = 0$ and $P \in R^1$. Then since $I + S = \text{const.}$, say 1, the system reduces to one equation

$$\frac{\partial I}{\partial t} = (1 - I) \int_{-\infty}^{\infty} k(P - P')I(t, P')\, dP'.$$

Let us suppose k satisfies the equation

$$d^2k/dP^2 = \gamma k(P) - \varepsilon\delta(P),$$

where γ and ε are positive. This condition means that k is an exponentially decaying function of P as $P \to \pm\infty$ whose derivative has a jump at $P = 0$ of order ε. Then setting

$$w = \int_{-\infty}^{\infty} k(P - P')I(t, P')\,dP',$$

we see that w satisfies

$$\frac{\partial^2 w}{\partial P^2} = \gamma w - \varepsilon I.$$

The system thus becomes

$$\partial I/\partial t = (1 - I)w,$$

$$\partial^2 w/\partial P^2 = \gamma w - \varepsilon I.$$

As before, we analyze this system for travelling wave solutions

$$I(P - ct), \quad w(P - ct).$$

Substitution of these functions into the equation gives

$$\text{(III.10)} \qquad -cI'(\xi) = (1 - I)w, \quad w''(\xi) = w - I,$$

where we have written $\xi = P - ct$ and $' = d/d\xi$.

Mollison establishes the following theorem.

THEOREM (Mollison 1972)). *Suppose the conditions listed above hold. Then there is a critical value c_0 such that* (III.10) *has a unique orbit with* $0 \leqq w, I \leqq 1$ *for each* $c \geqq c_0$.

The proof of Mollison's results involves analysis of some topological properties of the trajectories defined by (III.10) and it will not be presented here. However, a similar, though easier, analysis is carried out in the next section for the diffusion approximation.

(e) *Progressing wave solutions for one-dimensional diffusion approximation.* The diffusion approximation derived in (b), but with $P = x \in R^1$ and α a fixed constant can be written as

$$\text{(III.11)} \qquad \begin{aligned} \partial S/\partial t &= -S[\alpha I + \beta\partial^2 I/\partial x^2], \\ \partial I/\partial t &= S[\alpha I + \beta\,\partial^2 I/\partial x^2] - qI, \\ \partial Q/\partial t &= qI, \end{aligned}$$

where $-\infty < x < \infty$. The analysis of progressing wave solutions given here is quite similar to that of Kolmogoroff, Petrovsky and Piscounov (1936) and Kendall (1965).

The third equation in (III.11) is neglected, and we look for solutions of the remaining equations for S and I which are functions of the new variable

$$\xi = x - ct.$$

The parameter c denotes the velocity of the wave, and it remains to be determined Substituting into the differential equation, we have

$$cS' = S(\alpha I + \beta I''), \quad cI' = qI - S(\alpha I + \beta I''), \tag{III.12}$$

where $' = d/d\xi$ This system can be simplified somewhat. Integrating the first equation over some interval between a and ξ gives

$$c \log [S/S(a)] = \int_a^\xi I + \beta[I' - I'(a)].$$

The second equation gives

$$c[I + S - I(a) - S(a)] = q \int_a^\xi I$$

Combining these two equations, we have

$$\beta I' = \beta I'(a) + c \log (S/S(a)) - (\alpha c/q)[I + S - I(a) - S(a)].$$

Differentiating this equation and substituting the result in the first equation in (III.12) gives

$$S' = (q/c)I - (1/\beta)\{\beta I'(c) + c \log (S/S(a)) - (\alpha c/q)[I + S - I(a) - S(a)]\}.$$

These calculations have reduced the original problem (III.12) to a system of first order equations which contain certain new constants. We shall assume that far to the right there are no infectives. Thus, we assume

$$S(\infty) = S_0, \quad I(\infty) = I'(\infty) = 0.$$

Then allowing $a \to \infty$, we have

$$\begin{aligned} S' &= (q/c)I - (c/\beta) \log (S/S_0) + (\alpha c/q\beta)(I + S - S_0), \\ I' &= (\alpha c/q\beta)(S_0 - I - S) + (c/\beta) \log (S/S_0). \end{aligned} \tag{III.13}$$

Long before the wave arrives at a point, the state of the population is given by $S = S_0$. Long after the wave has passed, the number of susceptibles is given by s_0 ($\geqq 0$) and the population then is assumed to consist only of susceptibles and removals.

In order that there be a steady state $S = s_0$, $I = 0$, we must have

$$\log (s_0/S_0) = (\alpha/q)(s_0 - S_0) \quad \text{or} \quad s_0 = S_0 \exp [(\alpha/q)(s_0 - S_0)].$$

This second equation, rewritten with $F = s_0/S_0$, is equivalent to (III.7), and we have seen that there is a solution for F with $F < 1$ if and only if

$$S_0\alpha/q > 1.$$

Thus, we assume that at each point the initial susceptible population is above its threshold value.

The main results of this section are summarized next

THEOREM. *Let S_0, α, β and q be positive constants with $S_0 > (q/\alpha)$. Then there is a number.*

$$c_{\min} = 2\sqrt{\beta S_0[q - \alpha S_0]}.$$

such that for each $c \geqq c_{\min}$, (III.13) defines a unique trajectory $S_c(\xi)$, $I_c(\xi)$, joining $(s_0, 0)$ and $(S_0, 0)$ with $0 \leqq S_c \leqq S_0$ and $0 \leqq I_c$. Also, for $c < c_{\min}$, there is no such trajectory.

The theorem shows that for each $c > c_{\min}$, there is a unique (up to translation) progressing wave solution admissible for our problem in the sense that

$$S(\infty) = S_0, \quad I(\infty) = 0, \quad S(-\infty) = s_0, \quad I(-\infty) = 0, \quad \text{and}$$

$$I \geqq 0, \quad 0 \leqq S \leqq S_0$$

That is, for any velocity $c \geqq c_{\min}$, there is a unique admissible progressing wave which propagates with velocity c. Moreover, there is no admissible progressing wave with velocity of propagation less than $c_{\min}$.

Also, if the condition

$$\alpha S_0 > q$$

is not satisfied, there is no admissible progressing wave possible. This condition is again interpreted as a lower bound on the population density of susceptibles for them to support a progressing epidemic.

The stability of these epidemic waves can be analyzed just as the gene waves were in Part II

Finally, the proof of this result is based on the ideas used by Kolmogoroff et al. for the genetics model.

The theorem is proved in the following way. The SI-plane is shown in Fig. III.4. Under the conditions of the theorem, there are only two rest points in the (closed) first quadrant. $(s_0, 0)$ and $(S_0, 0)$ The first can be shown to be a saddle point with unique trajectory τ emanating from it into the first quadrant The rest point $(S_0, 0)$ is a stable point, however. for $c < c_{\min}$ it is a spiral point. Thus, any solution nearing it must eventually have its S component larger than S_0, and so be inadmissible. For $c \geqq c_{\min}$, a straight line, l, passing through $(S_0, 0)$ with negative slope can be found which solutions of (III.13) cannot cross from below. Finally, for $I \gg 0$, $I' < 0$. and so the region a is invariant to solutions It follows from the Poincaré-Bendixson theory that τ must converge to $(S_0, 0)$.

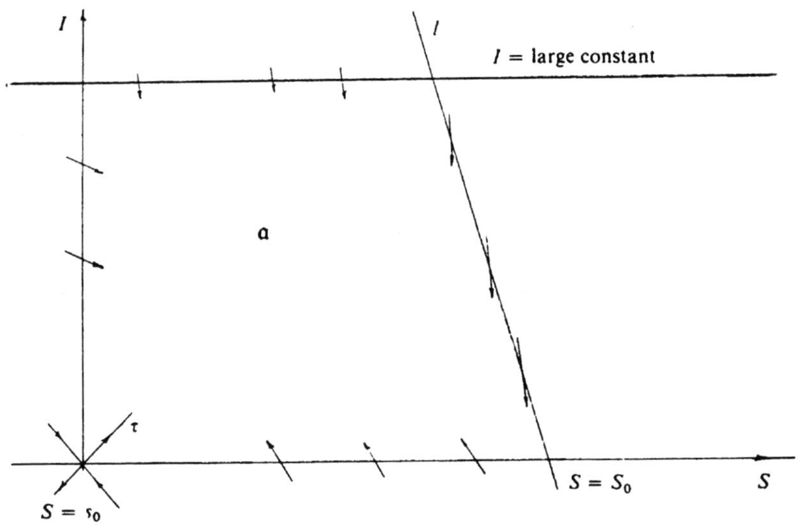

FIG III 4

BIBLIOGRAPHY

N. T. J. BAILEY (1957) *The Mathematical Theory of Epidemics*, Griffin, London.

R. BELLMAN AND K. COOKE (1963) *Differential-Difference Equations*, Academic Press, New York.

K. L COOKE (1967) *Functional differential equations. Some models and perturbation problems*, Differential Equations and Dynamical Systems, Hale and LaSalle, eds., Academic Press, New York.

K L. COOKE AND J A YORKE (1973) *Some equations modelling growth processes and gonorrhea epidemics*, Math Biosci., 16, pp. 75–101

R A. FISHER (1936) *The wave of advance of advantageous genes*, Ann. Eugen., 7, pp. 355–369.

J M GREENBERG AND F. HOPPENSTEADT (1975) *Asymptotic behavior of solutions to a population equation*, SIAM J Appl. Math , 28, pp 662–674

H. HETHCOTE (1970) *Note on determining the limiting susceptible population in an epidemic model*, Math Biosci., 9, pp. 161–163.

F HOPPENSTEADT AND P. WALTMAN (1970, 1971) *A problem in the theory of epidemics, I, II*, Math Biosci., 9, pp. 71–91; 12, pp. 133–145

F HOPPENSTEADT (1974) *Thresholds for Deterministic Epidemics*, Lecture Notes in Biomathematics, no 2, P. von den Driessche, ed., Springer-Verlag, New York.

——— (1974b) *An age dependent epidemic model*, J Franklin Inst., 297, pp 325–333

D G. KENDALL (1965) *Mathematical models of the spread of infection*, Math. and Comput Sci. in Biology and Medicine, H.M Stationery Office, pp 213–225.

W O. KERMACK AND A G. MCKENDRICK (1927, 1932, 1933) Proc. Roy Soc A, 115, pp. 700–721; 138, pp. 55–83; 141, pp. 94–122

H G LANDAU AND A RAPOPORT (1953) *Contribution to the mathematical theory of contagion and spread of information*, Bull. Math Biophys , 15, pp. 173–183.

D LUDWIG (1974) *Stochastic Population Theories*, Lecture Notes in Biomathematics, no 3, Springer-Verlag, New York

G MACDONALD (1965) *The dynamics of helminth infections, with special reference to schistosomes*, Trans R S Top. Med. Hyg . 59, pp 489–506.

H MARCHAND (1956) *Essai d'etude mathematique d'une forme d'epidemie*, Ann. U. Lyon. A 3, pp 13–46

D. MOLLISON (1971) *Spatial propagation of simple epidemics*, Thesis, King's College, Cambridge.

I NASELL AND W M. HIRSCH (1973) *The transmission dynamics of schisotosomiasmis*, Comm Pure Appl. Math., 26, pp. 395–453

H. E. SOPER (1929) *Interpretation of periodicity in disease prevalence*, J. R Statist. Soc. B, 92, pp. 34–73.

D STIRZAKER (1974) *A perturbation analysis for the epidemic with super infection*, Math. Biosci., in press

——— (1974b) *On the stochastic recurrence of epidemics*, J. Inst. Math. Applics., in press.

P WALTMAN (1974) *Deterministic Threshold Models in the Theory of Epidemics*, Lecture Notes in Biomathematics, no. 1, Springer-Verlag, New York

E B WILSON AND M H. BURKE (1942) *The epidemic curve*, Proc. Nat. Acad. Sci., 28, pp. 361–366.

L O. WILSON (1972) *An epidemic model involving a threshold*, Math. Biosci., 15, pp. 109–121.

K YOSIDA (1960) *Lectures on Differential and Integral Equations*, Interscience, New York.